Digital Libraries - Methods and Applications

Digital Libraries - Methods and Applications

Editor

Dominika Dechiel

scitus
academics

Digital Libraries - Methods and Applications
Edited by **Dominika Dechiel**

ISBN: 978-1-68117-242-2
Library of Congress Control Number: 2016934782

www.scitusacademics.com
Box No. 4766, 616 Corporate Way,
Suite 2, Valley Cottage,
NY 10989

Notice

Preface

Digital library is commonly seen as a type of information retrieval system which stores and accesses digital content remotely via computer networks. A digital library is a special library with a focused collection of digital objects that can include text, visual material, audio material, video material, stored as electronic media formats (as opposed to print, microform, or other media), along with means for organizing, storing, and retrieving the files and media contained in the library collection. Digital libraries can vary immensely in size and scope, and can be maintained by individuals, organizations, or affiliated with established physical library buildings or institutions, or with academic institutions. The digital content may be stored locally, or accessed remotely via computer networks. An electronic library is a type of information retrieval system.Digital Libraries acting as knowledge depositories can be considered as complex information systems, development and use of which require solution of numerous scientific, technological, methodological, economic, legal and other issues. Digital library technologies are rapidly developing. Challenges in semantics, integration of information, and perceptions of presentation of various kinds of data call for significant innovations. Development of Digital library technologies is becoming more and more essential for raising the standards of health care, education, science and economy, as well as the quality of life on the whole.Digital libraries are more and more actively coming into use within scientific organizations and universities. This book,Digital Libraries - Methods and Applications, is an endeavour to share the practical experiences of solutions to the operation of digital libraries. To indicate interdisciplinary routes to successful applications, this book explores the implication of digital libraries from the viewpoints of design, operation, and promotion.

Table of Contents

1

CHAPTER

A FULL TEXT RETRIEVAL SYSTEM IN A DIGITAL LIBRARY ENVIRONMENT

Kehinde Daniel Aruleba[1], Dipo Theophilus Akomolafe[2*], Babajide Afeni[3]

[1]*Department of Mathematics & Computer Science, Elizade University, Ilara-Mokin, Nigeria*
[2]*Department of Mathematical Sciences, Ondo State University of Science and Technology, Okitipupa, Nigeria*
[3]*Department of Computer Science, Joseph Ayo Babalola University, Ikeji-Arakeji, Nigeria*

ABSTRACT

The volume of information being created, generated and stored is huge. Without adequate knowledge of Information Retrieval (IR) methods, the retrieval process for information would be cumbersome and frustrating. Studies have further revealed that IR methods are essential in information centres (for example, Digital Library environment) for storage and retrieval of information. Therefore, with more than one billion people accessing the Internet, and millions of queries being issued on a daily basis, modern Web search engines are facing a problem of daunting scale. The main problem associated with the existing search engines is how to avoid irrelevant information retrieval and to retrieve the relevant ones. In this study, the existing system of library retrieval was studied. Problems associated with them were analyzed in order to address this problem. The concept of existing information retrieval models was studied, and the knowledge gained was used to design a digital library information retrieval system. It was successfully implemented using a real life data. The need for a continuous evaluation of the IR methods for effective and efficient full text retrieval system was recommended.

Keywords: Full text, Information Retrieval, Library, Digital Library, Queries, Indexing, Catalogue

1. INTRODUCTION

For Centuries, libraries have been organizing reading materials on shelves for easy access. However, systematic methods that had been widely adopted for the organization of library materials and their recordings for use by readers came into being a little more than a century ago [1] . The term digital library is used to refer to a library where some or all of the holdings are available in electronic form, and the services of the library are also made available electronically-frequently over the Internet so that users can access them remotely [2] . The primary purpose of digital libraries is to enable searching of electronic collections distributed across networks, rather than merely creating electronic repositories from digitized physical materials.

An information retrieval (IR) system is designed to retrieve any documents or information required by the user community. It is primarily targeted to make the right information available to the right user at right time. IR is concerned with representing, searching, and manipulating large collections of electronic text data. IR is a discipline that deals with retrieval of unstructured data or partially structured data, especially textual documents, in response to a set of query or topic statement(s), which may itself be unstructured [3] . IR system does not inform i.e. change the knowledge of the user on the subject of his enquiry; it merely informs the user of the existence or non-existence and whereabouts of documents relating to the request.

Many problems are associated with the current system of IR and such can be seen from the inability of the system to process request timely and to present inadequate results among others. In view of these inadequacies, it is imperative to develop an IR system that will curtail these inadequacies.

2. RELATED WORK

The importance of IR keeps growing as the amount of digital information keeps expanding at an ever-increasing rate. Stored documents, photographs and contents of books, and billions of Web pages are useful only if they can be easily found when needed.

2.1. Information Retrieval Models

For effectively retrieving relevant documents by IR strategies, the documents are typically transformed into a suitable representation. Each retrieval strategy incorporates a specific model for its document representation purposes. According to [4] , the Boolean model is the first model of IR and probably also the most criticized model. Larson [5] shows that much of this criticism seems to be based on lack of knowledge about how to utilise its search possibilities. In this model, we can pose any query which is in the form of a Boolean expression of terms, that is, in which terms are combined with the operators AND, OR, and NOT. The model views each document as just a set of words.

The vector space model (VSM) represents documents and queries as vectors in multidimensional space, whose dimensions are the terms used to build an index to represent the documents. It is used in IR, indexing and relevancy rankings and can be used in evaluation of Web search engines. According to Shang [6] , the VSM procedure was divided into three stages. The first stage is the document indexing where content bearing terms are extracted from the document text. The second stage is the weighting of the indexed terms to enhance retrieval of document relevant to the user. The last stage ranks the document with respect to the query according to a similarity measure.

According to Gonzalez [7] , Language models (LM) for information retrieval are retrieval models (taken from the speech recognition field) that do not impose an explicit parametric form for the probability of relevance. Lafferty and Zhai [8] presented a formal connection between probabilistic and language models. The basic idea of the language modelling approach to IR is to assume that a query Q is generated by a probabilistic model of document D. In this context, the generative language models approach estimate £¡ is the probability of the query being generated by a document.

2.2. Query Types

There are many different ways of searching for information. Here we describe the most common ones according to Salerma [9] .

A normal query is any query that is not explicitly indicated by the user to be a specialized query. For queries containing only a single term, the desired semantics are clear: match all documents that contain the term. For multiword queries, however, the desired semantics are not so clear. Some implementations treat it as an implicit Boolean query by inserting hidden AND operators between each search term.

Phrase queries are used to find documents that contain the given words in the given order. Usually phrase search is indicated by surrounding the sentence fragment in quotes in the query string. They are most useful for finding documents with common words used in a very specific way [9] . For example, if you do not remember the author of a paper, searching for it on the Internet as a phrase query will in all likelihood find it for you.

Boolean queries are queries where the search terms are connected to each other using the various operators available in Boolean logic, most common ones are AND, OR and NOT [10] . Usually parentheses can be used to group search terms. A simple example is software AND database, and a more complex one is software AND database AND data structure.

3. INFORMATION RETRIEVAL AND DIGITAL LIBRARIES

Libraries have been in existence since the beginning of writing and have served as a repository of the intellectual wealth of society. As such, libraries have always been concerned with storing and retrieving information in the media it is created on. As the quantities of information grew exponentially, libraries were forced to make maximum use of IR methods to facilitate the storage and retrieval process. Some of the IR methods used in digital libraries are described in the following:

Indexing: IR systems need an indexing mechanism for performing efficiently the retrieval process [7] . Indexing is the transformation from the received item to the searchable data structure. Building an index from a document collection involves several steps, from gathering and identifying the actual documents to generating the final data structures [11] .

Catalogue records: are short records that provide summary information about a library object. The word catalogue is applied to records that have a consistent structure, organized according to systematic rules. Library catalogues serve many functions, not only information retrieval. Some catalogues provide comprehensive bibliographic information that cannot be derived directly from the objects. This includes information about authors or the provenance of museum artefacts [12] . Descriptive Metadata: Many methods of information discovery do not search the actual objects in the collections, but work from descriptive metadata about the objects [12] . The metadata typically consists of a catalogue or indexing record, or an abstract, one record for each object. Usually it is stored separately from the objects that it describes, but sometimes it is embedded in the objects. Descriptive metadata is usually expressed as text, but can be used to describe information that is in formats other than text, such as images, sound recording, maps, computer programs, and other non-text materials, as well as for textual documents.

Library Digitization

According to Ian and David [13] , defined digitization as the process of taking traditional library materials that are in form of books and converting them to the electronic form where they can be stored and manipulated by a computer.

According to Alhaji [14] , there are three main reasons for digitization of a library system

1) To make the documents more accessible: This is to serve existing library users better, i.e. to allow users search the full text of documents or to allow users search from remote locations.
2) To preserve the documents: Allow user read older or unique documents without damage to the originals
3) To reuse the documents: Allowing conversion of documents into different formats.

4. METHODOLOGY

From the architecture of a FTRS described in Aruleba et al. [15] , a full text search retrieval system was designed. This section presents the modelling of the system. The modelling is in two parts, which are: Analysis and Design

The analysis of the existing information system in University of Ilorin library was extensively carried out by studying the existing environment. The result of the analysis is presented using the Use-case diagram shown in figure 1.

The result of the existing library information shows that the entire system is made of seven steps. The first step is where the potential library user is registered. The registration allows the user to become a registered and legally authorised user of the library. After the registration, the user is allowed to use the facilities provided by the library. After registration, the registered user is allowed to undertake the remaining steps that is registered user can check available books, read books, check document title, author, publisher, borrow book, return book and pay fine in case of late submission of book(s).

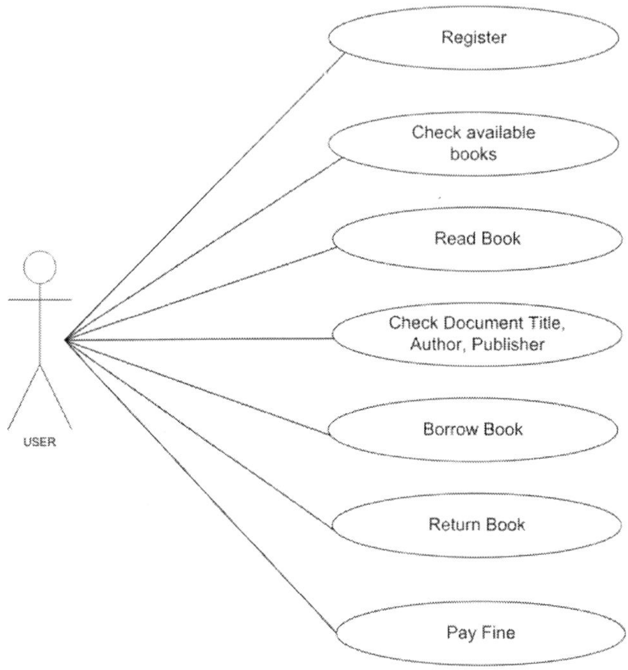

Figure 1. The result of analysis of UNILORIN library system.

The proposed system in addition to the functionality of the existing system allows users to search, modify user details, and upload documents as shown with use-case in Figure 2.

From Figure 2, the proposed system is made up of eight distinct steps. Though the components are interwoven, each of them performs distinct functions but all work together as a system to process request timely.

4.1. Database Design

Database design mainly includes requirement analysis, concept structure design stage, the logic structure design stage, physical structure design stage, database implementation stage, database operation and maintenance stage, there are six steps altogether.

From the analysis done, Table 1 was designed for the implementation of the proposed system.

4.2. User Interface Design

There are many factors that must be considered when designing the user interface of a software because the user must be able to interact with the system in a way that the system will understand whatever input given by the user. Therefore, the quality of the interface and software in general must pass the usability testing standard. Some usability factors, such as fit for use, ease of learning, task efficiency, ease to remember, subjective satisfaction and understand ability but all are put into consideration when designing the user interface (Figure 3).

The home page screen depicted in figure 4, contains four major modules which are the Search, Registration, Request and Login while the Admin module home page shown in figure 5, contains Sub-module which are view students, view staff, view books, create new book, view book request, create/view facilities, create/view department, logout. Each of them will lead you to its database when clicked and manipulated.

5. SYSTEM IMPLEMENTATION PHASE AND TESTING

This phase implements what have been discussed in thesection 4. The system was developed and implemented with PHP and MySql Technology.

Table 1. Generated database.

Table	Action
Admin	To authenticate the system user
Book request	To store information about requested books by the system users
Books	To store book details
Department	To store all the departments available
Faculty	To store all the faculties available
Staff	To store information of staffs using the system
Students	To store information of students using the system.

Figure 2. Use case diagram showing the proposed system.

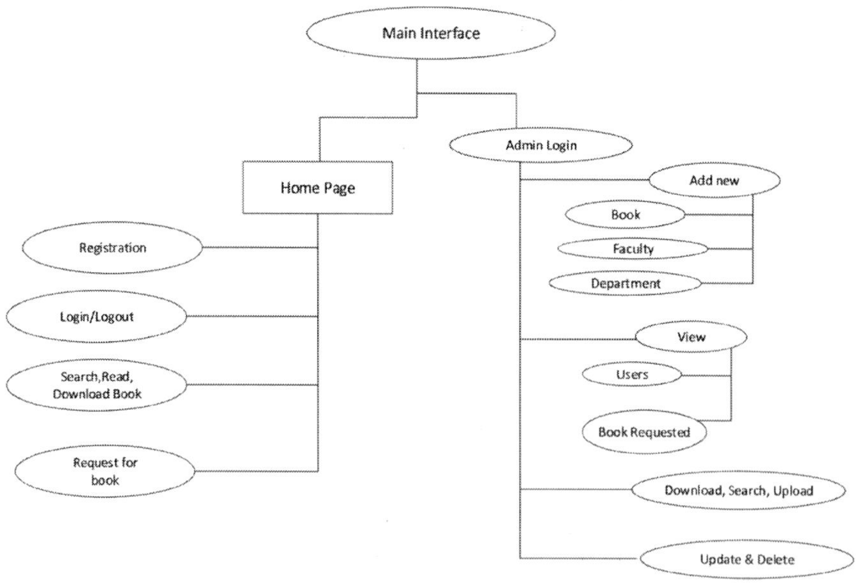

Figure 3. Showing the main interfaces of the system.

Figure 4. Home page design interface.

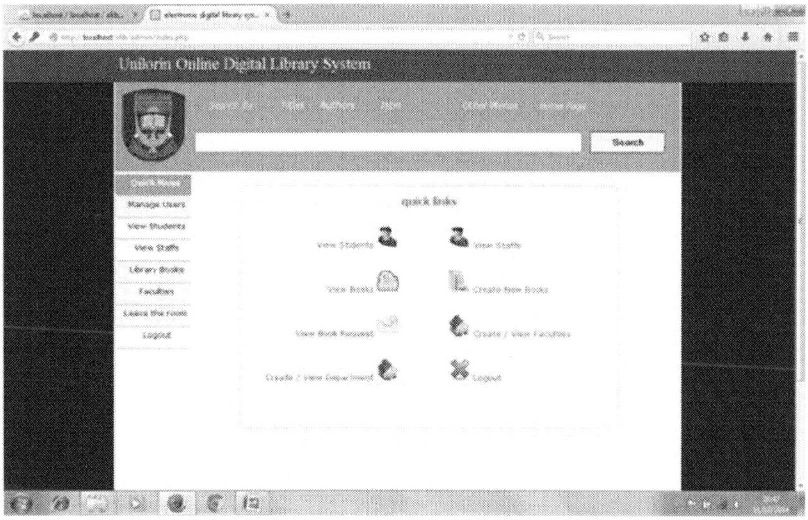

Figure 5. Admin home page design interface.

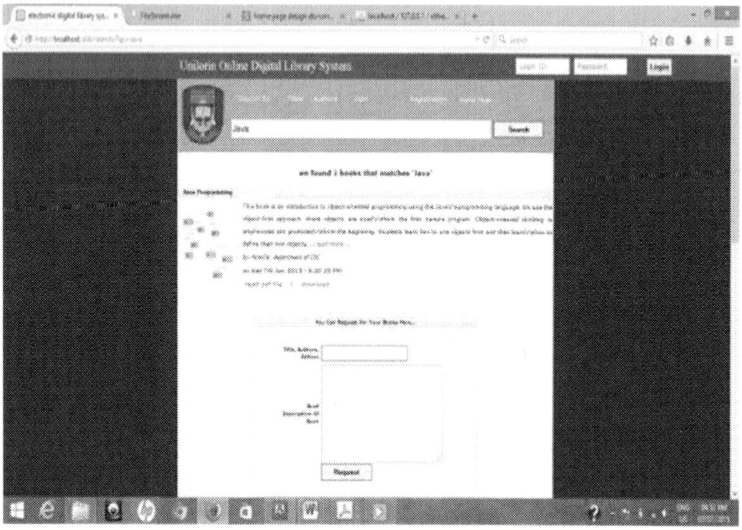

Figure 6. Home page implementation output.

Home Page Interface Implementation

The home page shown in figure 6 is the key aspect of the system, because it gives the basic user interface for the full text retrieval digital library. It comprises of: Search, Login, Registration and Request described as follows:

Search: This feature can be used by any user. This module provides a convenient book searching function, the user could search books based on a variety of conditions.

Login: Every user who wants to use the system is authenticated by means of username and password. All entered parameters of the password are matched with information stored in the database, therefore only authenticated users can log on to the program with limited access.

If the login information is wrong, the user will be notified of login failure and would need to try again.

Registration: This involves registering new users. It contains registration form interface with entries like email address, last name, first name, password, password confirmation and sex.

Request: If a user can find the specific book needed, request can be made for such book.

6. CONCLUSIONS AND FUTURE WORK

This study has been able to develop and successfully implement an IR system that reduces the hurdles associated with present system of searching in Libraries. It is shown in the system that users searching full text are more likely to find relevant articles than searching only abstracts. This finding affirms the value of full text collections for text retrieval and provides a platform for aligning searching algorithms that take advantage of rapidly-growing digital archives.

Also, the following areas of the study can be improved upon in future studies to create a more robust IR:

- Increase in the size of the database, this will enable large data storage.
- Integrate advert plans for research materials the institution wants to be selling online.
- Acquire and publish video, audio and heavy graphic research materials.

REFERENCES

1. Onwuchekwa, E.O. and Jegede, O.R. (2011) Information Retrieval Methods in Libraries and Information Centers. An International Multidisciplinary Journal, 5, Serial No. 23
2. Rosenberg, D. (2005) Towards the Digital Library: Findings of an Investigation to Establish the Current Status of University Libraries in Africa. International Network for the Availability of Scientific Publications, Oxford.
3. Greengrass, E. (2002) Information Retrieval: A Survey by Ed Greengrass. Information Retrieval, 141-163.
4. Goker, A. and Davies, J. (2009) Information Retrieval: Searching in the 21st Century. Wiley, West Sussex.
5. arson, R.R. (2010) Information Retrieval Systems. In: Bates, M.J. and Maack, M.N., Eds., Encyclopedia of Library and Information Sciences, 3rd Edition, CRC Press, New York, IV, 2553-2563.
6. Li, L.Z. and Shang, Y. (2000) A New Statistical Method for Performance Evaluation of Search Engines. ICTAI.

7. González, R.B. (2008) Index Compression for Information Retrieval Systems. Unpublished PhD Thesis, University of A Coruna, A Coruna.
8. Lafferty, J. and Zhai, C. (2003) Probabilistic Relevance Models Based on Document and Query Generation. In: Croft, W.B. and Lafferty, J., Eds., Language Modelling for Information Retrieval, Kluwer, Pittsburgh, 11-56.
9. Salerma, O. (2006) Design of a Full Text Search Index for a Database Management System. MSc Dissertation, University of Helsinki, Helsinki.
10. Pazer, J.W. (2013) The Importance of the Boolean Search Query in Social Media Monitoring Tools. Dragon Search.
11. Manning, C.D., Raghavan, P. and Schutze, H. (2008) Introduction to Information Retrieval. Cambridge University Press, Cambridge.
12. Arms, W. (2002) Manuscript of Digital Libraries. MIT Press, Cambridge.
13. Written, L.H. and Brainbridge, D. (2003) How to Build a Digital Library. Morgan Kaufman Publishers, London.
14. Alhaji, I. (2005) Digitization of Library Resources and the Formation of Digital Libraries: A Practical Approach. University of Pretoria, Pretoria.
15. Aruleba, K.D., Aremu, D.R., Oriogun, P.K., Agbele, K.K. and Agho, A.O. (2015) Evaluation of Full Text Search Retrieval System. Nigeria Computer Society, 26, 154-159.

2

CHAPTER

EXPLORING DIGITAL LIBRARIES THROUGH VISUAL INTERFACES

Beomjin Kim[1], Jon Scott[1] and SeungEun Kim[2]

[1]*Department of Computer Science, Indiana University-Purdue University Fort Wayne, USA*

[2]*Department of Multimedia Engineering, Seoul Women's University, South Korea*

INTRODUCTION

Libraries are long standing institutions, providing an important service of making information widely available. So it is with Digital Libraries (DL), but this evolution into a computerized format does not come without its own unique challenges. The variety and quantity of information available in the digital space is truly astounding. However, as this growth continues, traditional methods for searching are becoming less effective to support the needs of users to find information quickly and easily.

The conventional library book search system provides several attributes associated with books as a response to the users' inquiry. This includes title, author, publication year, ISBN, page total, and similar information. While considering the increasing volume of data, the current text-based approach to result presentation is not an ideal solution for the modern digital environment. Particularly in the case of comparing a lengthy list of search results, this approach is ill-suited, as it is inefficient and non-intuitive (Good et al., 2005). Assistance, such as ranked results, can aid in such problems but the user will still be relied on to investigate the top results individually (Veerasamy & Heikes, 1997; Dushay, 2004). Additionally, the current popular approach of presenting a summary of content may not accurately reflect what is of value to the user.

The advancements and trends that allow for the rapid growth of DL also permit more elaborate interfaces with which to access them (Bertini et al., 2005). Information Visualization is one such avenue, which has proven to be an effective approach in acquiring information from a large compilation of data. By making use of users' perceptual cognition for navigating extensive digital workspaces, their ability to understand, and speed with which they review the information space is improved (Card et al., 1999). Previous studies have proven the significance of visualization in the users' information forage (Veerasamy & Heikes, 1997; Hawkins, 1999; Kim et al., 2002).

One common approach to assist users' search activities uses visualization techniques combined with information filtering. The users' interaction defines the attributes of interest that easily filter out unrelated data. The following visualization procedure transforms the remaining data into graphical illustrations. FilmFinder and HomeFinder are interactive visual interfaces which assist the user to narrow down the search scope and easily compare results using the visualization (Ahlberg & Shneiderman, 1994; Williamson & Shneiderman, 1992). Another common approach focuses on presenting the underlying information through visualization. A novel text visualization interface, TileBars, shows the distribution behavior of a set of query terms (Hearst, 1995). This presentation allows users to compare multiple documents compactly and concurrently. Other visualization techniques in this category concentrate on showing a portion of the information at a great level of detail while maintaining the overall structure of the information (Lamping et al., 1995). Information clustering is another method for supporting navigation of a large data collection. Related documents are clustered together, whose notable characteristics are visualized using a mixture of attributes (Shneiderman et al., 2000; Au et al., 2000; Nowell et al., 1996). Visualization methodologies, such as those mentioned here, have shown that the illustration of data has contributed to improving the user's ability to comprehend information quickly. This, in turn, leads to increased speed and accuracy in finding desired information (Card et al., 1999; Kim, 2004).

We can apply these kinds of visual abstractions to enhance searches on different types of data domains. By presenting information in such a way, a large amount of content can be displayed in a format which is more intuitive. This has the benefit of allowing the user to analyze data more effectively, increasing the user's ability to comprehend results and make better content selections. Periscope, for example, is a visualization system targeted at web search results. It provides a series of different visualizations which users can utilize to analyze and explore the result set. A holistic interface can be used to organize documents into various categories, such as language and format, or web related attributes such as DNS domains. An analytical interface allows for up to seven attributes to be relayed at a time, through use of X, Y, Z axes, color, size, shape, and animation (Wiza et al., 2004).

The search based on the underlying content will increase the accuracy in finding targeted information from the available resources. Think for a moment how one might search a physical book for a topic of interest. The logical place to start would be the index. Indexes are valuable resources for referencing major terms which appear in a book. The categorical and hierarchical layout of terms in the index allows us to identify associated topics easily, along with their relationships without reading the underlying contents. The page numbers coupled with each entry makes it possible to estimate the amount information relating to a particular subject. The index will represent the overall layout of entire book contents. Due to a lack of readily available sources, this valuable information has been under utilized. The current trend of digitizing books in recent years allows us to exploit content in searches, instead of just depending on superficial book attributes. The visualization techniques utilizing this information will further enhances the user's search on the DL system.

The main objective of this book chapter is on the utilization of visualization techniques for exploring the DL system. The following chapter will survey and summarize various visualization approaches which applied to DL. We will introduce a visual interface which presents general book information through iconic representations. This chapter also proposes a novel visualization which utilizes the book index for mimicking the content analysis. It will allow for detailed comparison of index-based information between selected works. Two different visualizations for this detail view are implemented, each with different strengths. The procedure and analysis of results for a usability test follow, along with discussion and future possibilities, and final conclusions.

dynamic hyperbolic tree to show hierarchical relationships among documents. By allowing users to manipulate the manner in which data is displayed, these visualizations provided a better chance to find patterns within the data that may not typically be apparent.

Lin proposed another approach through a graphical table of contents (GTOC) that tries to exploit a different perspective of underling information by utilizing the table of contents of the book. GTOC showed the dimension of items in the table of contents based on Kohonen's self organizing feature map algorithm (Lin, 1996). The paper introduces how documents can be organized and then visualized to allow the user easy access of underlying contents. The GTOC prototype describes various interactive tools to assist the user exploring document contents and analyzing relationships among terms in the table of contents.

Both the attributes associated with documents and its underlying contents are valuable resources finding the relevant information of the users' interest. The increasing computing power and performance of graphics devices make it possible to exercise these information in presenting search results. The following section introduces a newly developed visualization system that will assist the user's search while utilizing the book index, which has been underutilized as a visualization resource.

METHOD

The Visual Interface for Digital Library Search (VIDLS) system utilizes an Overview+Detail approach for presenting book search results. This is a visualization technique that uses multiple images to display the entire data space, as well as show an up-close, detailed view of the data (Baudich et al., 2002; Shneiderman, 1996). Similar to traditional library searches, the overview will present outline of the book search results through graphical illustration. The user interactively selects a subset of visualized icons that will allow them to execute content level exploration. When a user provides search terms of interests, the Detail view presents a visualization which relates, in depth, the information presented by the index. For our pilot study, we developed two possible visualizations for this view of the index, each with its own strengths. One emphasizes space utilization, whereas the other is designed to focus on clearly presenting term relationships.

3.1 Overview visualization

The Overview allows the user to perform a general search on the data space, similar to traditional library tools. This visualization utilizes a tabular layout which offers a familiar spread-sheet style organization of book search results. The attribute of each axis can be independently selected by the user, allowing for a more targeted display and increasing the users' comprehension of the data set (Shneiderman et al., 2000). This functionality will assist the user in customizing the search based on their own judgment of which attributes are more important. In the Overview, each book is represented as a circular node, located in the appropriate cell based on the axes. To deliver the estimated amount of content, node size is determined by the normalized page count of the book compared to the rest. Books with a greater number of pages naturally map to the largest nodes. This will allow users to quickly identify the amount and distribution of content available.

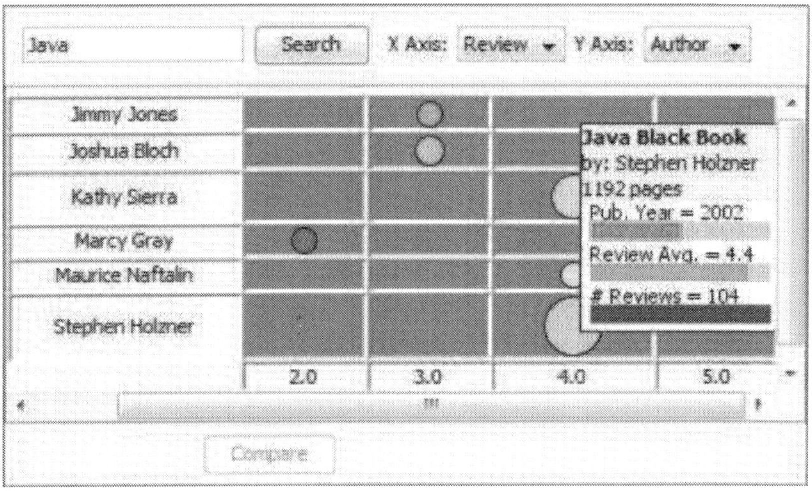

Figure 1. Overview interface displaying search results with tool-tip

The system presents other book attributes through a color coding scheme. The green component is derived through collaborative filtering. This is a content-filtering technique based upon the opinion of users whom have already evaluated the item in question (Resnick et al., 1994). This will be a valuable piece of information to know regarding the quality of the book. In the case of VIDLS, collaborative filtering would be done by collecting user reviews, much like you found find at a merchant website.

Higher intensity of the green component would indicate a more positive response from reviews.

The accuracy of the content-filtering is highly correlated with the number of evaluators and their preferences. As an example: the mean score of a book which has a low number of reviews could potentially be misleading and unreliable (Allen, 1990). To account for this, the Overview associates the blue component with the number of unique evaluations given to a book. The publication year is a valuable attribute to find the most up to date information. This metric has been represented through the red component. With the utilization of the RGB color scheme, the larger more recent books with a solid review will be closer to white in intensity, while older, smaller, poorly reviewed works tend toward black. If the user needs to know more precise information for the components, or other detailed information of the work, the system provides this via a tooltip interface (figure 1).

Detailed visualization

Once a user makes a set of selections from the Overview, the system will provide a detailed view of those items. This display is tasked with presenting visualized index data for each book, allowing the user to perform more refined queries, and compare and contrast each work in detail. Two visualizations were implemented on a data model to present the same information while emphasizing different attributes: one a radial graph, and the other a sunburst-like display (Stasko et al., 2000).

1.2.1 Data model

The underlying model for the detailed visualizations relies on several important characteristics associated with a term in the index. These are: number of pages, number of occurrences, content density, and location relative to the parent term. Although this information is readily available in the index, it is unintuitive in a text format. Presenting these attributes via graphical illustration, the user should gain greater understanding of the content, and make more informed choices of which books to choose.

The number of pages associated with a term by the index is a potential indicator as to how much coverage a topic receives in a book. A work with a greater number of pages on a subject is intuitively going to have more potential value than one with less. Naturally, this attribute is represented by size in the visualizations. To assist the user in quickly making general comparison between each book, this attribute is further expanded to be a cumulative page total of the term along with its sub-terms.

The number of occurrences and content density are related through a color code. This is intended to give the user a better understanding of the comparative value of each term being displayed. Term occurrence within the book is related by the green component. Although one work may have more information referenced in a single index entry for a topic, that does not necessarily make it the better choice. If another book contained references of a particular term in many different locations within the index, that may be an indication of a greater breadth or complexity of coverage on that topic. The number of term occurrences are normalized across the display, with the instances of greatest coverage mapping to the highest green intensity, and the one with the fewest having no green intensity.

Although two books may have the same total of pages for a particular topic, this should not be taken to mean they have equal value. Consider the situation in which one index allocates ten pages for a topic, but they are separate and not listed as a single range of pages. Another index containing the same topic, but listed as a single range of ten pages, such as 152-161 may be more valuable to the user. The first may have fragmented references of the topic, but no detailed coverage, whereas the second may have a full section or chapter dedicated to the subject. To relate this information to the user, the content density is mapped to the blue color component. The value is given by calculating the ratio of page continuity to the total number of pages for a term. This is shown by equation 1, where *Cib* is the blue intensity for term i, $\sum_{j=0}^{n} P_j^c$ is the number of individual page ranges for *i*, *Ti* is the total number of pages for the term *i*, and *Imax* is the brightest intensity of blue possible on the machine. This gives terms with more concentrated information a stronger representation in the visualization.

$$C_{ib} = \left(1 - \left(\frac{\sum_{j=0}^{n} P_j^c}{T_i}\right)\right) * I_{max} \tag{1}$$

The last attribute represents the relationship between a sub-term and its parent. Similar to the previous attribute, a sub-term which is located close to its parent within the book could be an indication of cohesive coverage of the topic. A user may regard such a case with more value than one where the sub-term is more removed from the parent's location. The value of this attribute is calculated by equation 2.

$$D_j = ABS\left(\frac{NORM\left(\sum_{k=0}^{n}(P_j^k - \overline{Root_i})\right)}{N}\right) \tag{2}$$

Dj is a normalized distance between the root node and associated sub-node *j*; P_j^k is the page number contains sub-term j;Rooti is the mean page number having the root word i; and NORM and ABS are a normalization and absolute function respectively.

1.2.2 Visualizations

The two visualizations share several aspects, in addition to the attributes described previously. Each displays the matching term of the search prominently in the center, with the sub-terms radiating outward. Since each term is sized based on the cumulative page count of itself and sub-terms, the root node will give users a quick guidance as to which book has more content. Additionally, each graph has a context side-bar that indicates which term is currently moused-over, along with a histogram displaying its color components. The similarities end here for the two visualizations.

The first visualization approach employs a radial graph layout; with each term presented as a circular node branching off the parent. Sub-terms are sized in the same way as the root, with node size directly correlated to the total number of pages. The largest sub-terms start from the top toward the bottom of the visualization. This is done to prevent over-crowding of the display. Terms with the fewest pages are considered less important, and are collapsed into a single node, indicated by a red outline, if not enough space is available. Another way in which space is conserved is by displaying only the first level of sub-terms. A node with hidden sub-terms has a ridged outline. These along with the red 'overflow' node can be clicked show the hidden terms (figure 2).

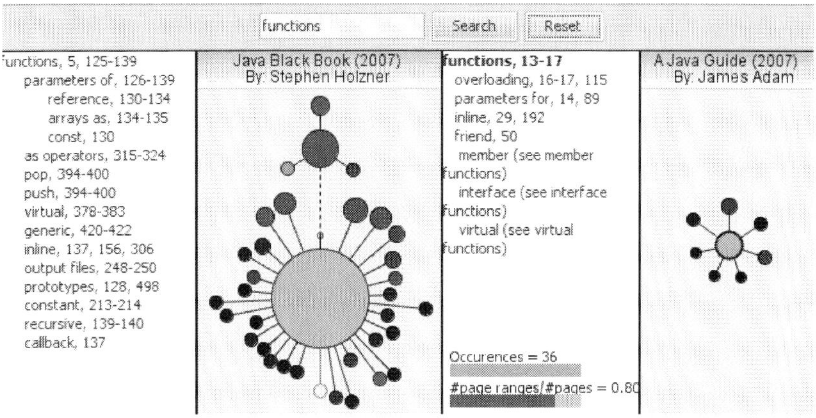

Figure 2. Diagram shows two books being compared in the Detail View. The left shows expanded sub-terms and an overflow node. The right has context highlighting of text list (bold-face font) and bar display from user interaction

In contrast, the second visualization follows a sunburst-like design. Sub-terms are displayed in concentric sections around the center, extending from their parent term. Each sub-term 'slice' is sized as a percentage of its page allocation from the parent's. This has the benefit that all terms can be displayed compactly without hiding any. No additional user interaction would be required to explore the full term hierarchy. One potential downside of this design, compared to the previous one, is that terms are crowded together. This may cause difficulty in picking them out from their neighbors.

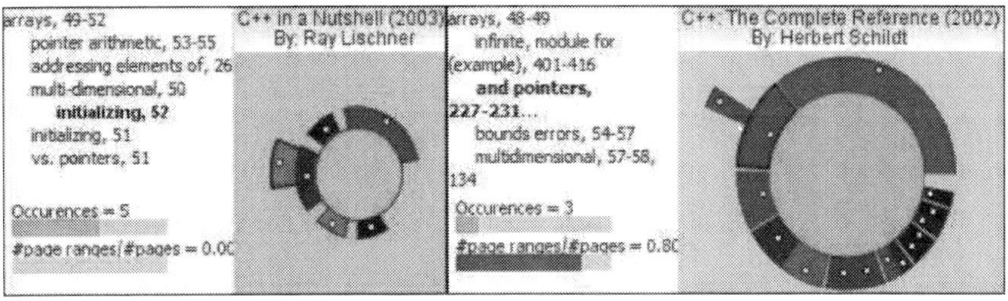

Figure 3. Alternate sunburst detailed view

In the radial graph, the relation between a sub-term and its parent is indicated by the branch length. Terms which appear on pages near those of the parent naturally are drawn with shorter branches. Since the sunburst approach has forgone branches to conserve space, a different analogue is required. Each slice instead contains a bubble, which will relate this information (figure 3). As the closeness to parent is normalized, the term with the closest relationship has a bubble at the base of the slice, while the term with the most distant has a bubble on the outer edge of its slice.

As was previously noted, the chief difference between the two visualizations is their concern for space conservation. With its emphasis on the term relation branch, and distinct nodes, the radial graph may have more clarity, but at the cost of poorer space utilization. This results in the need to hide terms considered less important. The other visualization puts more value in making better use of space, and is able to display all terms efficiently. However, when many terms are displayed at once, the resultant crowding could make smaller slices difficult to see. The addition of the bubble overlay to each slice can add to this difficulty.

EXPERIMENT

To measure the effectiveness of VIDLS, a series of usability tests were performed to compare it to more traditional text-based interfaces. In this set of experiments the VIDLS system, using the radial graph detailed layout, was used in conjunction with a text-only interface for the same information.

Each usability test consisted of a small group of three to five students, comfortable in the use of computers, whom had no prior experience with VIDLS. Each group was given a brief orientation presentation, and then allowed to acclimate to the functionality of both applications for a few minutes. The tests then began, with each user asked to search for books and information with the text-only application and then with the visualization based approach. Once the test concluded, participants then filled out a survey to measure their quantitative and qualitative feedback of VIDLS in comparison to the text-based format.

The questionnaire has 14 questions which utilize a 5 point Likert scale; 5 indicating the highest level of satisfaction, and 1 the lowest. Table 1 shows the survey results that summarizes the user's feedback to VIDLS over a text-based library search system. Participants overall found the VIDLS system to be satisfactory, with some comments highlighting that unfamiliarity with the visualizations led to preference of the text-based design.

The Overview visualization was well received, with 82% of testers responding that it improved their ability to quickly identify desirable books. This was supported through the exploratory aspects of the system, which participants indicated positively as giving meaningful result displays, and facilitating better understanding of the information.

The radial graph visualization of index level information was also seen in a positive light, as 65% found it to be a useful representation, and only 11% preferring the familiar text-only listing of results. As with the Overview, testers considered the Detail View to also be useful in expressing the concentration of information, and improving selection among the results. Table 1 shows a summary of survey results.

Table 1. Post experiment survey results

Question	Pos.	Neut.	Neg.
VIDLS overall was preferable to a text-based search	53%	35%	12%
The Overview improved identification of desirable books	82%	6%	12%
Use of the selectable axis facilitated a better understanding of a set of books.	71%	24%	5%
The Overview helped in selecting a subset of books.	65%	29%	6%
It was easy to discern book attributes based on node color	53%	18%	29%
The Detailed view was preferable to a text approach	65%	24%	11%
It was easy to discern term attributes based on node color	47%	24%	29%
The Detail view made relevant book selection easier	88%	12%	0%
The relation between a term and it's child was understandable	65%	35%	0%
The Detail view helped identify terms related to the search	94%	6%	0%

DISCUSSION AND FUTURE WORK

Overall, feedback was positive toward the VIDLS system. Users indicated it was a helpful and effective alternative to more traditional search utilities. Among the responses, familiarity with text-based result displays was a common explanation for favoring it over a new approach. It is promising that the majority of participants still showed preference for the visualized prototype.

One of the chief strengths of VIDLS highlighted by the usability tests is its exploratory nature. The customization afforded by the selectable axes in the overview, and the interactive nature of the detailed view greatly enhanced users search activity. Many felt that their understanding of the information was improved by these traits. This aspect of the system is also important, in that it may help lead users to other terminology related to their goal, but exempt from their initial search vocabulary.

Although most were comfortable with it, the color-coded attributes proved to be the main area of difficulty for users. The primary issue highlighted by the experiment is user difficulty with interpreting the color codes of both views. Although around half in both instances were comfortable with this aspect of the system, around 30% had trouble with it. The post-experiment interviews provided two main causes for the diverging opinions. First, the RBG color model was not familiar to some. These users cited heavy exposure to the RYB (primary color pigments) model as being a source of confusion when interpreting the displays. The increased unfamiliarity left those individuals feeling more comfortable using a text-based search.

The other difficulty reported by users was in determining the relative value of one result with another. This could be, in part, a result of the human eye being more sensitive to some colors rather than others. For example, green-yellow colors have the strongest reception, which could mislead a user into considering a result with this color to have more overall value than another when that may not be the case (Foley et al., 1996). Researching and examining alternate color models will be one of the challenging task for future work on the VIDLS system.

While users found the graph approach to be effective, it has plenty of room for improvement. The issue of overlapping prevents showing all information at once. If possible, it would be preferable to avoid that situation, as it may reduce the effectiveness of the visualization. As stated earlier, the layout does give the graph design more clarity and makes it easy to pick out individual items; however some users found it difficult to compare two nodes which aren't adjacent to each other.

The sunburst layout could improve upon these points. The compact design greatly reduces the need to hide information for typical indexes without extensive hierarchies. It may also cut down on unnecessary competing visual information created by the empty space seen in the graph layout. Another possibility is that the slice shape of sub-term nodes is more distinct, and may afford more visual context when making comparisons.

Though small in size, the experiment supported the viability of VIDLS as a search utility. Still, larger experiments will need to be planned in order to better understand how the prototype compares against existing search methodologies. Testing will also need to be done to confirm how the sunburst and radial visualizations differ in effectiveness. These will be key topics as future work on VIDLS continues.Other areas for future work include investigating content analysis. Although indexes proved to be a useful foundation for visualization, these are limited to books. Other media lack such precompiled information. VIDLS would need to integrate methods to examine document content in order to extract the information it presents. This is a very ambitious topic, but could expose additional attributes for expressing to the user, as well as allowing for a more universal application of the system. Similarly, enhancing VIDLS to have thesaurus-like capability to identify similar topics would be another valuable addition. Grouping such items together would further improve a users understanding of which work has more comprehensive coverage on a topic.

CONCLUSION

The emergence of the Digital Library provides an interesting dichotomy: effortless access to a large compilation of data, but increased challenges in finding a few specific works of interest. Having effective mechanisms for exploring this space is both an important and also demanding research topic. Information visualization can be one possible solution in addressing these difficulties presented by digital library systems. This chapter surveyed existing approaches, applying a range of visualization techniques for querying information in the digital library system with a focus toward books. The developed interfaces presents multiple attributes associated with books through visual icons on a table. This simple graphical illustration assists the user to compare and contrast search results intuitively. The chapter also introduces a novel method that utilizes book indexes for enhancing searches. The exploitation of indexes will further enhance search activities, making them more efficient and effective. The conducted survey-based usability tests that compared the proposed system over traditional text-based approaches showed the efficacy of the visual interfaces as a search supporting tool on digital libraries. This chapter concludes with the belief that visualizations will be a promising approach to address current issues in the digital library system suffering from the complexity and volume of its sources.

REFERENCES

1. Abbasi, A. & Chen, H. (2007). Categorization and Analysis of Text in Computer Mediated Communication Archives Using Visualization, *Proceedings of the 7th ACM/IEEE-CS Joint Conference on Digital Libraries*, pp.11-18
2. Ahlberg, C. & Shneiderman, B. (1994). Visual information seeking using the FilmFinder, *Proceedings of the ACM Conference on Human Factors in Computing Systems*, pp.433-434
3. Allen, R. (1990). User Models: Theory, Method, and Practice, *International Journal of Man- Machine Studies*, Vol. 32, pp. 511-543
4. Au, P.; Cary, M.; Sewraz, S.; Guo, Y. & Rüger, S. (2000). New Paradigms in Information Visualization, *Proceedings of the 23rd Annual International ACM SIGIR Conference on Research and Development in Information Retrieval*, pp. 307-309
5. Baudich, P.; Good, N.; Bellotti, V. & Schraedley, P. (2002). Keeping Things in Context: A Comparative Evaluation of Focus Plus Context Screens, Overviews and Zooming, *Proceedings of the SIGCHI Conference on Human Factors in Computing Systems*, pp. 259- 266
6. Bederson, B.; Shneiderman, B. & Wattenberg, M. (2002). Ordered and Quantum Treemaps: Making Effective Use of 2D Space to Display Hierarchies, *ACM Transactions on Graphics*, Vol. 21, No. 4, pp. 833-854
7. Bertini, E.; Catarci, T.; Di Bello, L. & Kimani, S. (2005). Visualization in Digital Libraries, *Lecture Notes in Computer Science*, Vol. 3379, pp. 183-169, Springer-Verlag, Berlin
8. Borner, K. & Chen, C. (2002). Visual Interfaces to Digital Libraries: Motivation, Utilization, and Socio-technical Challenges, *Lectures Notes in Computer Science*, Vol. 2539, pp. 1-9 Springer-Verlag, Berlin Heidelberg New York

9. Card, S.; Mackinlay, J. & Shneiderman, B. (1999). *Readings in Information Visualization Using Vision to Think*, Morgan Kaufman
10. Chen, C. (1999). Visualizing Semantic Spaces and Author Co-citation Networks in Digital Libraries, *Information Processing and Management: an International Journal,* Vol. 35, No. 3, pp. 401-420
11. Christel, M. & Martin, D (1998). Information Visualization Within a Digital Video Library, *Journal of Intelligent Information Systems,* Vol. 11, No. 3, pp. 235-257
12. Clarkson, E.; Desai, K. & Foley, J. (2009). ResultMaps: Visualization for Search Interfaces, *IEEE Transactions on Visualization and Computer Graphics,* Vol. 15, No. 6, pp. 1057-1064
13. Dushay, N. (2004). Visualizing Bibliographic Metadata - A Virtual (Book) Spine Viewer. *D- Lib Magazine,* Vol.10, No. 10
14. Good, L.; Popat, A.; Janssen, W. & Bier, E. (2005). Fluid Interface for Personal Digital Libraries, *Proceedings of the 9th European Conference on Research and Advanced Technology for Digital Libraries,* pp. 162-173
15. Hawkins, D. (1999). Information Visualization: Don't Tell Me, Show Me!, *Online,* Vol. 23, No. 1, pp. 88-90
16. Hearst, M.A. (1995). Tilebars: Visualization of term distribution information in full text information access. *Proceedings of the ACM Conference on Human Factors in Computing Systems,* pp. 59-66
17. Kampanya, N.; Shen, R.; Kim, S.; North, C. & Fox, E. (2004). Citiviz: A Visual User Interface to the CITIDEL System, *Lecture Notes in Computer Science,* Vol. 3232, pp. 122-133, Springer-Verlag, Berlin
18. Kim, B.; Johnson, P. & Huarng, A. (2002). Colored-sketch of Text Information, *Journal of Informing Science,* Vol. 5, No. 4, pp. 163-173
19. Kim, B. (2004). Visual Interface for Evaluating Internet Search Results, *Lecture Notes in Computer Science,* Vol. 2973, pp. 533-542, Springer-Verlag, Berlin
20. Lamping, J.; Rao, R. & Pirolli, P. (1995). A Focus+Context Technique Based on Hyperbolic Geometry for Visualizing Large Hierarchies, *Proceedings of the ACM Conference on Human Factors in Computing Systems,* pp. 401-408

21. Lin, X (1996). Graphical Table of Contents, *Proceedings of the 1st ACM International Conference on Digital Libraries*, pp.45-53
22. Marks, L.; Hussel, J.; McMahon, T. & Luce, R. (1996). ActiveGraph: A Digital Library Visualization Tool, *International Journal on Digital Libraries*, Vol. 5, No. 1, pp. 57-69
23. Nowell, L.; France, R.; Hix, D.; Heath, L. & Fox, E. (1996). Visualizing Search Results: Some Alternatives to Query-Document Similarity, *Proceedings of the 19th Annual International ACM SIGIR Conference on Research and Development in Information Retrieval*, pp. 67-75
24. Nowell, L.; France, R. & Hix, D. (1997). Exploring Search Results with Envision, *Proceedings of the ACM SIGCHI*, pp. 14-15
25. Resnick, P.; Iacovou, N.; Suchak, M.; Bergstrom, P. & Riedl, J. (1994). GroupLens: An Open Architecture for Collaborative Filtering of Netnews, *Proceedings of the ACM Conference on Computer Supported Cooperative Work*, pp. 175-186
26. Shen, R.; Vemuri, N.; Fan, W.; Torres, R. & Fox, E. (2006). Exploring Digital Libraries: Integrating Browsing, Searching, and Visualization, *Proceedings of the 6th ACM/IEEE-CS Joint Conference on Digital Libraries*, pp. 1-10
27. Shiaw, H.; Jacob, R. & Crane, G. (2004). The 3D Vase Museum: A New Approach to Context in a Digital Library, *Proceedings of the 4th ACM/IEEE-CS Joint Conference on Digital Libraries*, pp. 125-134
28. Shneiderman, B (1996). The Eyes Have It: A Task by Data Type Taxonomy for Information Visualizations, *Proceedings of the IEEE Symposium on Visual Languages*, pp.336-343
29. Shneiderman, B.; Feldman, D.; Rose, A. & Grau, X. (2000). Visualizing Digital Library Search Results with Categorical and Hierarchical Axes, *Proceedings of the 5th ACM Conference on Digital Libraries*, pp. 57-66
30. Silva, N.; Sánchez, A.; Proal, C. & Redbollar, C. (2003). Visual Exploration of Large Collections in Digital Libraries, *Proceedings of the Latin American Conference on Human-Computer Interaction*, pp.147-157
31. Stasko, J.; Catrambone, R.; Guzdial, M. & McDonald, K. (2000). An Evaluation of Space-Filling Information Visualizations for Depicting Hierarchical Structures, *International Journal of Human-Computer Studies*, Vol. 53, No. 5, pp. 663-694

32. Foley, J.; Van Dam, A.; Feiner, S. & Hughes, J. (1996) *Computer Graphics: Principles and Practice*, 2nd ed. Addison-Wesley Publishing Company
33. Veerasamy, A. & Heikes, R. (1997). Effectiveness of a Graphical Display of Retrieval Results, *Proceedings of the 20th Annual International ACM SIGIR Conference on Research and Development in Information Retrieval*, pp. 236-245
34. Williamson, C. & Shneiderman, B. (1992). The dynamic HomeFinder: evaluating dynamic queries in a real-estate information exploration system, *Proceedings of the 15th Annual International ACM SIGIR Conference on Research and Development in Information Retrieval*, pp. 338-346

3
CHAPTER

KNOWLEDGE AND SKILLS FOR THE DIGITAL ERA ACADEMIC LIBRARY

J. Raju *

Library and Information Studies Centre, University of Cape Town, Private Bag X3, Rondebosch, 7701 South Africa

ABSTRACT

Technology has altered the traditional academic library beyond recognition. These dramatic changes have impacted significantly on the knowledge and skills requirements for LIS professionals practising in this environment. While there have been studies in other parts of the world which have investigated the knowledge and skills requirements for the digital era academic library environment, to date no comprehensive study has 'drilled' down into this area in the South African context. This paper reports on a preliminary study which is part of a wider study aimed at developing a comprehensive skills statement which would provide an objective framework against which professional LIS practitioners in the modern academic library environment in South Africa may both measure their existing competencies and also identify the need for further skills acquisition. The research question guiding this preliminary investigation was: What key knowledge and skills are required for LIS professionals to effectively and efficiently practise in a digital era academic library in South Africa? The triangulated findings (using content analysis of job advertisements and semi-structured interviews) from this preliminary investigation are used to ascertain an initial picture of key knowledge and skills sets required for LIS professionals in this environment. These preliminary findings also proved useful in teasing out some of the parameters for the wider study targeting the development of a comprehensive skills statement for higher education libraries in South Africa. The study reported here has relevance for the academic library context in other parts of the world as well.

Keywords: Academic libraries; Digital era; Disciplinary knowledge; Generic skills; Job advertisements; Personal competencies

INTRODUCTION

New methods of scholarly communication, expansion of the library's virtual space via knowledge or research commons, the proliferation of social media, and the explosive growth of mobile devices, tablets and related applications, have collectively altered the traditional academic library beyond recognition. These dramatic changes, largely the result of rapidly evolving information and communication technologies (ICTs), have impacted significantly on the knowledge and skills requirements for library and information science (LIS) professionals practising in this environment. The transformed landscape requires a new generation of LIS professionals to effectively and efficiently mediate it. Orme (2008, p. 627–628) categorised knowledge and skills required for this transformed environment into: discipline-specific knowledge (that is, knowledge that relates specifically to the LIS profession), generic skills (general skills which apply to all disciplines) and personal competencies (attitudes, values and personal traits). Choi and Rasmussen (2009, p. 465), through content analysis of job advertisements in the United States of America (USA), found that key disciplinary knowledge required for this digitally oriented environment included understanding metadata, and knowledge and experience in digital content creation and management. Generic skills such as effective communication and interpersonal skills, critical thinking, problem solving and teamwork were found by Nonthacumjane (2011, p. 283) to be required by information professionals in a digital library environment in both Norway and Thailand. Howard's (2009) Australian master's study (cited in Nonthacumjane, 2011, p. 283) highlighted personal competencies such as flexibility, adaptability and reflective thinking as being required for working in a digital library environment. What about knowledge and skills requirements for the digital era South African higher education academic library environment? To date, no comprehensive study has 'drilled' down into this area in the South African context. South Africa is richly endowed with academic libraries. These are located in the country's 23 universities. Many of these higher education libraries offer state-of-the art LIS services to academic and research communities in universities in the country which are leading institutions on the African continent. Hence an analysis of knowledge and skills requirements for the digital era academic library environment in South Africa is important.

This paper reports on a preliminary study which is part of a wider study aimed at developing a comprehensive skills statement which would provide an objective framework against which professional LIS practitioners in the modern academic library environment in South Africa may both measure their existing competencies and also identify the need for further skills acquisition. Such a skills statement would also be useful in informing curriculum review and revision in LIS education and training, as academic libraries in South Africa are a major employer of LIS graduates (Ocholla & Shongwe, 2013, p. 38). There are ten LIS schools in South Africa, based in universities, which offer education and training in response to the knowledge and skills requirements of the LIS work environment. The research question guiding this preliminary investigation was: What key knowledge and skills are required for LIS professionals to effectively and efficiently practise in a digital era academic library in South Africa? The triangulated findings (using content analysis of job advertisements and semi-structured interviews) from this preliminary investigation are used to ascertain an initial picture of key knowledge and skills sets required for LIS professionals in this environment. These preliminary findings also proved useful in teasing out some of the parameters for the wider study targeting the development of a comprehensive skills statement for higher education libraries in South Africa. While these early findings draw from South African higher education libraries, the study is contextualized within global trends affecting higher education libraries and hence has relevance for other parts of the world as well.

Literature review

A review of pertinent literature locates this study in the global context of the changing academic library and its impact on knowledge and skills requirements for academic libraries in the digital era.

The academic library in the digital era

As technology continues to impact on the delivery of information services, traditional academic library systems, point out Choi and Rasmussen, 2006 and Choi and Rasmussen, 2009, have come to embrace the digital library model. Academic libraries have evolved from focusing on the management of physical resources and related services to "transforming resources and services into digital formats to support teaching, learning

While '...technology associated with LIS application...' is treated as a separate trend in Table 1 (for purposes of honing in on this significant trend), in other parts of this paper, as explained earlier, it is discussed under disciplinary skills due to the close association between digital library applications and ICTs. Table 1 also provides percentages (calculated against the 11 core research papers selected by the researcher for inclusion in the table) reflecting relative preponderance in the literature of the identified trends. Table 1 proves useful, in the Discussion section of this paper, in effecting comparisons between the findings of the preliminary study being reported and that in the literature reviewed in this paper.

METHODOLOGY

The preliminary study reported here adopts a qualitative approach by using content analysis of job advertisements and semi-structured interviews with purposively selected academic library professional LIS personnel to preliminarily ascertain the knowledge and skills requirements for a modern academic library in South Africa. As mentioned at the outset, the research question guiding this preliminary investigation was: What key knowledge and skills are required for LIS professionals to effectively and efficiently practise in a digital era academic library in South Africa? Content analysis is a commonly used descriptive technique for analysing the content of a document or other communication to discover features and patterns (Neuman, 2006, p. 44). Content analysis of job advertisements is a well-established method of researching requirements of the employment market in a particular sector (Orme, 2008, p. 620, p. 623) and is very useful in reflecting the demands of employers, employment opportunities and emerging trends in the employment market.

In view of the fact that this study was a precursor to an envisaged wider study, the *Mail & Guardian*, a weekly newspaper known nationally for carrying advertisements for vacancies in the higher education sector in South Africa, was initially used as a single source of advertisements for professional LIS posts (that is, posts requiring a professional LIS qualification) in academic libraries in South Africa. The researcher (towards the end of 2012) attempted to source relevant advertisements from the popular 'Jobs and Academic' section of the newspaper made available to the researcher at the newspaper's Cape Town satellite office. The researcher began searching, retrospectively, for advertisements from 2008 onwards on the assumption that this would deliver a significant number of 'recent' advertisements for LIS professional positions in academic libraries.

The *Mail & Guardian* did not have a publically available electronic means of searching its content — hence the retrospective searching had to be done manually. The researcher did not have access to issues of the newspaper that were at the binders at the time of searching. Added to this was the fact that a more recent trend is that many institutions do not place the full advertisement in the press — often an abbreviated form, largely for reasons of costs, appears with a URL to access the full advertisement which is removed from the relevant website once the closing date has passed, thus 'blocking' access to important research data. Reeves and Hahn (2010, p, 118) also allude to this difficulty: "... job ads typically appear now online, ...a distinct advantage for job-seekers, but a serious problem for future studies of this type if the ads are not captured and stored before they disappear". The net result of the retrospective searching for advertisements was that only 39 were found. The researcher supplemented this by searching differently for 2013 advertisements. The *Mail & Guardian* was purchased on a weekly basis and scanned immediately for relevant advertisements, including searching websites referred to by abbreviated advertisements. Over and above this, *LiasaOnline*, which is the listserv of the Library and Information Association of South Africa (LIASA), was accessed for relevant advertisements, following the advice that "...online job ads are essential to getting a comprehensive picture of the job market (Reeves & Hahn, 2010, p. 105)". This approach for the whole of 2013 yielded a further 32 advertisements bringing the total to 71 (excluding duplicate) advertisements.

To triangulate findings from the job advertisements for this exploratory exercise, the researcher purposively selected, for conducting semi-structured interviews, three professional LIS staff members each (six interviews in total) from the academic libraries of the University of Cape Town and Stellenbosch University, two of South Africa's research-led universities with world rankings — an obvious indicator of their state-of-the art academic libraries. It was also the intention of the researcher to use this modest number of interviews to trial this particular method of data collection for this type of study.

The interviewees in the purposive sampling were selected based on the researcher's knowledge and research experience (Welman, Kruger, & Mitchell, 2005, p. 69) with the identified research sites (the two academic libraries). This purposive selection included individuals occupying positions such as Senior Librarians (2) of branch library collections (Medicine and Law), Manager (1) of a disciplinary information service

6. Luce, R. E. (2008). A newvalue equation challenge: The emergence of eResearch and roles for research libraries. Retrieved January 22, 2013, from. http://www.clir.org/pubs/ reports/pub142/luce.html
7. McCarthy, J. (2005). Planning a future workforce: An Australian perspective. New Review of Academic Librarianship, 11(1), 41–56.
8. Middleton, M. (2003). Skills expectations of library graduates. New Library World, 104(1184/1185), 42–56.
9. Missingham, R. (2006). Library and information science: Skills for twenty-first century professionals. Library Management, 27(4/5), 257–268.
10. Neuman, W. L. (2006). Social research methods: Qualitative and quantitative approaches. Boston: Pearson.
11. Nonthacumjane, P. (2011). Key skills and competencies of a newgeneration of LIS professionals. IFLA Journal, 37(4), 280–288.
12. O'Connor, S., & Au, L. -C. (2008). Steering a future through scenarios: Into the academic library of the future. The Journal of Academic Librarianship, 35(1), 57–64.
13. Ocholla, D., & Shongwe, M. (2013). An analysis of the library and information science (LIS) job market in South Africa. South African Journal of Libraries and Information Science, 79(1), 35–43, http://dx.doi.org/10.7553/79-1-113.
14. Orme, V. (2008). You will be ...: A study of job advertisements to determine employers' requirements for LIS professionals in the UK in 2007. Library Review, 57(8), 619–633.
15. Partridge, H., & Hallam, G. (2004). The double helix: A personal account of the discovery of the structure of [the information professional's] DNA. Paper presented at the ALIA 2004 Biennial Conference, Gold Coast, Australia, 21–24 September 2004 Retrieved January 25, 2013, from http://conferences.alia.org.au/alia2004/pdfs/ partridge.h.paper.pdf.
16. Reeves, R. K., & Hahn, T. B. (2010). Job advertisements for recent graduates: Advising, curriculum, and job-seeking implications. Journal of Education for Library and Information Science, 51(2), 103–119.
17. Riley-Huff, D. A., & Rholes, J. M. (2011). Librarians and technology skill acquisition: Issues and perspectives. Information Technology in Libraries, 30(3), 129–140.

18. Tammaro, A.M. (2007). A curriculum for digital librarians: A reflection on the European debate. New Library World, 108, 229–246.
19. University of Melbourne (2008). Future skills for academic librarians: research and research and training paper for Information Futures Commission. Retrieved March 15, 2012, from. www.informationfutures.unimelb.edu.au/_data/assets/pdf_file/0010/105103/15-research20080509.pdf
20. Welman, C., Kruger, F., &Mitchell, B. (2005). Research methodology (3rd ed.). Cape Town:Oxford University Press.
21. Wise, S., Henninger, M., & Kennan, M.A. (2011). Changing trends in LIS job advertisements.
22. Australian Academic and Research Libraries, 42(4), 268–295.

4
CHAPTER

DEVELOPING AND USING A GUIDE TO ASSESS LEARNING RESOURCE QUALITY IN EDUCATIONAL DIGITAL LIBRARIES

Heather Leary[1], Sarah Giersch[2], Andrew Walker[1] and Mimi Recker[1]

[1] *Utah State University,*

[2] *Columbia University United States of America*

INTRODUCTION

The democratization of content creation via ubiquitous Internet tools and infrastructure (Anderson, 2006) has fueled an explosion of user-generated digital content in the commercial and educational markets. Federal agencies in the United States, such as the National Science Foundation, are actively seeking ways to integrate teachers and learners into the *education cyber-infrastructure* (Computing Research Association, 2005; Pea et al., 2008). The overall purpose of this education cyberinfrastructure is to connect people with information and with web-based tools to improve productivity, teaching, and learning. By connecting teachers and learners with tools, they become co-creators of educational content. Educational digital libraries, part of this cyberinfrastructure, have been created as places to deposit and disseminate educational content, which often takes the form of online learning resources of varying levels of granularity. Digital library content can be created easily and deposited rapidly, placing it outside the peer review processes typically employed by publishers and professional societies, but it is unclear whether teachers are using high quality

I see it." The problem (or the solution?) lies in the perspective of users who are confronted with a variety of resources of indeterminate origin. For school administrators, teachers, and learners, a high quality resource could consist of an entire lesson that meets several state standards; a web-based animation that accurately depicts a science concept; or, an essay that provides an answer in a quiz.

Grappling with the idea of "quality", several education digital libraries have already created rubrics to assess their content. As we tried to apply these criteria to the content in the Instructional Architect, as described below, we observed that portions of each rubric were applicable in different settings, and other portions were specific to each digital library with little room for re-use outside of the original context (Martin, 2004). As such, our goals were

1) to synthesize the various dimensions of existing rubrics in order to identify a standardized set of criteria that could potentially be used by any digital library with online educational resources (Giersch, Leary, Palmer & Recker, 2008a), and 2) to create a guide that could be used to assess the quality of materials in the Instructional Architect projects (http://ia.usu.edu). This chapter briefly describes our process for developing the standardized set of assessment criteria; evaluating its utility and usability with middle school science and math teachers; developing a Quality Guide for use with Instructional Architect projects; testing its reliability; and, exploring how the Guide could foster teachers' skills in designing learning resources.

PREVIOUS RESEARCH

Many education digital library builders have developed rubrics to help assess the quality of online educational resources generated by teachers, faculty and learners. However, the motivations and methods for implementing the rubrics vary as do the implied assumptions about the value of the results for each education digital library.

Established publishing houses and professional societies institute peer review processes to maintain a reputation for producing quality publications thereby increasing subscriptions and membership. For newly-funded education digital libraries, under pressure to prove their value, rubrics and a process to measure quality, were created to first *establish* a reputation with users (Robertshaw, Leary, Walker, Bloxham, Recker, 2009). Similarly, other education digital libraries used rubrics to support access to high quality resources in a digital library or repository (Liu & Ward, 2007; McMartin, 2004) because use of a collection development policy for inclusion of quality resources adds value and buy-in from peer reviewers and users (Sumner et al, 2003). Still other sites developed a rubric to guide authors in creating high quality online resources by gathering feedback from target users (McMartin, 2004).

The methods by which education digital libraries collected data using rubrics and how they used that information also varied. Muramatsu and Agogino (1999) used a rubric to gather feedback from *targeted users* to inform collection development, while Recker, Walker, and Lawless (2003) gathered data through automated mechanisms to provide teachers with suggestions based on resources they had previously viewed and used. Establishing peer- review panels and processes was not a *de facto* choice for every education digital library (Fitzgerald, Lovin, & Branch, 2003; ORC, McMartin, 2004). Enlisting users-as-builders, or reviewers-as-builders, is a characteristic of the grass-roots environment in which many education digital libraries evolved over the last decade in the U.S. However, many digital library developers found that gathering reviews from users was difficult, though providing incentives, in the form of awards (Muramatsu & Agogino, 1999) or other methods (McMartin, 2004), was one way to increase acceptance and adoption of the rubric and review process and of the education digital library in general.

The process of creating and testing an assessment rubric can be time consuming. Many rubric creators believe it is important to map out stakeholders and end-users, review previous work by other digital libraries (Knox et al, 1999), and learn how to gather data (Recker, Walker, Lawless, 2003; Sumner et al., 2003). Testing a rubric with users, for formative evaluation, further refines the rubric, yielding usability information (Fitzgerald, Lovin, & Branch, 2003) as well as real time feedback from target users (Recker, Walker, Lawless, 2003; Sumner et al., 2003), which ultimately assists with the process of implementing the rubric with a larger group of users.

CONTEXT

The Instructional Architect

The Instructional Architect is a simple, web-based authoring tool designed to help K-12 teachers find, design, and use online learning resources in their classrooms (Recker, 2006). When using the Instructional Architect, teachers search for and save links to online learning resources from the Web and educational digital libraries such as the National Science Digital Library (http://nsdl.org). Resources that can be linked include online content, RSS feeds and podcasts. Once resources are gathered users then create 'Instructional Architect projects' (in the form of web pages) that contain instructional objectives, activities, and assessments. The resources are either embedded into the project or link out to web-based resources. In this way, teachers create Instructional Architect projects that customize resources to their local context. Figure 1 shows an Instructional Architect project and an accompanying resource.

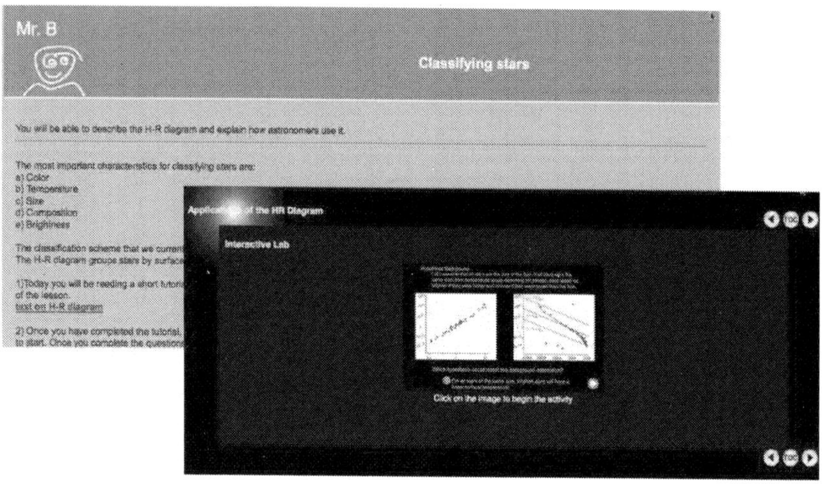

Figure 1. Example of an Instructional Architect project created by Mr. B and related resource.

Initially the majority of users were K-12 teachers in the U.S. who participated in professional development workshops from 2005-2009. Recently, the Instructional Architect system has garnered more widespread usage and spread 'virally' to teachers not directly attending these workshops. Table 1 shows statistics on Instructional Architect usage and growth in the number of Instructional Architect projects and resources over the last 12 months.

Table 1. Instructional Architect usage data (to August 2010).

Data	*N*	*12-month growth*
Registered users	5,300	23%
Instructional Architect projects created	11,500	37%
Online learning resources used	52,100	67%
Visits to Instructional Architect project (since 8/2006)	1 million	64%

Before any of the abundant Instructional Architect projects created by users can be ingested into educational digital libraries, like the National Science Digital Library, to become part of the education cyberinfrastructure, they need to be evaluated for quality. To support this outcome, the Instructional Architect projects must be reviewed with a rubric that is reliable across multiple sites, a process that is scalable across the thousands of IA projects, and that has useful results for end-users, reviewers, and digital library developers.

METHOD

Synthesizing existing rubrics & creating the Quality Guide

Following the recommendations above for establishing a rubric to assess content (Knox et al, 1999; Recker, Walker, Lawless, 2003; Sumner et al., 2003), we began with a literature review in the disciplines of computer science, library and information science, education (specifically online learning objects), and digital libraries. We selected only articles that included or referenced rubrics used to review online educational resources. The literature review yielded articles with descriptions of 12 rubrics (Liu, & Ward, 2007; Fitzgerald, Lovin,

& Branch, 2003; Nesbit, Belfer, Leacock, 2003; Ohio Resource Center; Knox et al., 1999; Recker, Walker, & Lawless, 2003; Sumner, Khoo, Recker, & Marlino, 2003; Custard & Sumner, 2005; Kastens, DeFelice, Devaul, DiLeonardo, Ginger, Larsen, et al., 2005; McMartin, 2004; Muramatsu & Agogino, 1999).

In keeping with our first goal (to synthesize the various dimensions of existing digital library rubrics in order to identify a standardized set

We then used a qualitative analysis to address research question two with the understanding that participants had learned how to use online educational resources. Using a constant comparative analysis (Glaser & Strauss, 1967), data from the reflection papers, discussion boards, and one question in the pre-and post-survey were mapped into themes and analyzed. Participants reported in the pre-survey that when evaluating online educational resources they looked for fit with the curriculum, accuracy, ease of use, currency, text readability, and recommendations by others. After completing this course module, participants added in their post-survey comments that they looked for: content quality, distractions on the resource pages, credibility of the site, and engagement. Many of the criteria they added were items listed in the Quality Guide. Thus, it appeared that use of the Quality Guide helped refine participants' approach to designing online learning resources, in the form of Instructional Architect projects.

To answer research question three, the comments from the discussion boards in Blackboard where participants posted their evaluations of their peers' Instructional Architect projects were used. Some participants commented on the readability and text clarity of the Instructional Architect project, and many caught spelling errors. One participant commented after receiving some feedback, "I have learned that having a peer read through my projects can be very valuable. They catch mistakes that I do not see." Overall, 53% of the participants reported that providing and receiving feedback from their peers via reviews was valuable.

Two additional themes surfaced from analyses of the discussion boards during the process of peer evaluation that related to re-use of other teachers' Instructional Architect projects and ideas. First, participants asked if they could use an Instructional Architect project they reviewed, or said they planned to use it in their classroom. The second theme was that participants learned what others had done and wanted to implement the same idea in their project. A participant commented that, "Completing a peer review gives us an opportunity to see other creations and improve our programs as we see fit."

The final data collection point included participants' reflection papers. Participants were asked to report on their instructional situation, successes and difficulties, what they learned from their peers and using the Quality Guide, and how they could improve their Instructional Architect project. Three areas of learning were repeatedly reported in the reflection papers: what they learned by reviewing peer work; what they learned for their own work; and, the value of re-using Instructional Architect projects. Similar ideas were also expressed throughout the discussion boards and survey answers.

Reliability and improvements

Version one (see Figure 3) of the Quality Guide was tested for reliability in order to answer research question four. Peer ratings of Instructional Architect projects using the Guide, provided by the students in the online graduate course, were used for this analysis. The Quality Guide criteria (5 stars) were scored on a scale of 0-4, where zero or one star is low (e.g., very unclear) and four or five stars is high (e.g., very clear). See Figure 3 above. Means and standard deviations were run for each question.

Criteria one through four and six of the Quality Guide were analyzed using an intra-class correlation (scale of -1 to 1, high agreement is found as it approaches 1, with less agreement as it approaches -1). Criteria five was not analyzed as it asks for a classification on what the Instructional Architect project is doing (e.g., teaching a new concept, reinforcing a concept, etc). Table 4 shows the intraclass correlation values for each question analyzed.

A close look at the data (see Table 3) shows that the teachers generally rated their peers at an average of 2 or more (3 or more stars on the Quality Guide) for any of the criteria on the Guide. This could possibly be attributed to the fact that educators as a whole tend to provide high ratings for their peer, a phenomenon observed in prior work (Walker, Recker, Lawless, & Wiley, 2004). As a result of the high ratings, the full range of the scale was not used, and slight departures were magnified in the intraclass correlation. The negative values indicate that there was more variation between projects than there was between raters. As is shown in the low intraclass correlation (Table 4), agreement on those ratings was never high, and in most cases was non-existent.

Table 3. Means and standard deviations for criteria 1-4, 6 of the Instructional Architect Quality Guide.

Criteria	*Mean*	*Standard Deviation*
1 – Content accuracy	2.50	1.13
2 – Text clarity	2.34	0.98
3 – Links in project	3.30	0.83
4 – Project completeness	2.14	1.50
6 – Overall rating	3.22	1.84

Table 4. Intraclass correlation (ICC) values for criteria 1-4, 6 of the Instructional Architect Quality Guide.

Criteria	*Intraclass Correlation*
1 – Content accuracy	0
2 – Text clarity	-0.129
3 – Links in project	0.031
4 – Project completeness	-0.89
6 – Overall rating	-0.129

The reliability analysis suggests that the scale needs more explanation for each item and that the scale should encompass either a broader range or a dichotomous rating (yes or no). Improvements made to the Quality Guide as a result of this testing included adding more detailed definitions of the review criteria, re-arranging the order of criteria, and modifying scoring instructions. As a result of this testing, improvements were made to create the final version of the Quality Guide (see Figure 4), which is currently used to inform and support individual teachers as they create Instructional Architect projects.

INSTRUCTIONAL ARCHITECT™

Use the following indicators to inform your choices when designing or reviewing IA projects.

Accuracy

Information is credible, truthful, reliable, current

Very inaccurate	*Inaccurate*	*Accurate*	*Very accurate*
Comments:			

Text Clarity

Appropriate grade level, length, and amount (as concise as needed)

Very unclear	*Unclear*	*Clear*	*Very clear*
Comments:			

Links in Project

Work and go to the correct page

All links fail	*Most links fail*	*Most links work*	*All links work*
Comments:			

Project Completeness

Includes the state standard, objectives, learning goal, assessment, or an example

Very Incomplete	*Incomplete*	*Complete*	*Very complete*
Comments:			

The Project

Provides a resource list	*Teaches a concept*	*Reinforces a concept*	*Provides a summary of content*	*Other*
Comments:				

Overall

Meets no criteria	*Meets some criteria*	*Meets all criteria*	*Exceeds criteria*
Comments:			

UtahState UNIVERSITY

Figure 4. Final version of the Instructional Architect Quality guide.

CONCLUSION

The foundational list of criteria (see Table 2), which was derived from rubrics of 12 different education digital libraries, can be used by other digital library developers to evaluate the quality of online learning resources, or it can be modified to fit the need of local contexts, such as the Instructional Architect. We created the Instructional Architect Quality Guide to encourage change in teachers' perspective and behaviors around designing and using online educational resources. The results of our evaluation indicate that participants found value in the Quality Guide as a means to improve their own projects through completing and receiving reviews. Unfortunately, we found that the Instructional Architect Quality Guide does not scale well when it is used to determine which of the 11,500 Instructional Architect projects are of sufficient quality to be ingested by the National Science Digital Library.

FUTURE RESEARCH

Future work is focused on addressing these scalability issues through using an already- developed and -tested automated and scalable quality assessment method, which will better support and facilitate teacher co-creation of online content. Similar to the work described in this chapter, this approach relies upon distilling the elusive notion of quality into a set of concrete indicators. We are also testing a previously developed machine learning algorithm, called Opera (Bethard et al., 2009), to assess whether its quality ratings along these indicators match teacher assessments (Recker et al., 2011). Results from these ongoing studies will help determine if Opera can serve as a proxy for the laborious and expensive task of having teachers or peer reviewers assess the quality of online resources and IA projects. We are still motivated by having online learning resources and Instructional Architect projects reviewed so that quality content can be ingested by educational digital libraries, such as the National Science Digital Library and ultimately included in the education cyberinfrastructure.

ACKNOWLEDGEMENTS

The authors would like to thank the study participants for their time and assistance, as well as members of the Instructional Architect research group. This material is based upon work supported by the National Science Foundation under Grant No. 0554440, and Utah State University. Any opinions, findings, and conclusions or recommendations expressed in this material are those of the authors and do not necessarily reflect the views of the National Science Foundation.

REFERENCES

1. Anderson, C. (2006). *The Long Tail: Why the future of business is selling less of more.* Hyperion, New York.
2. Bethard, S., Wetzler, P., Butcher, K., Martin, J. H., Sumner, T. (2009). Automatically Characterizing Resource Quality for Educational Digital Libraries. In *Proceedings of the 9th ACM/IEEE-CS Joint Conference on Digital Libraries.* New York: ACM.
3. Cohen, J. (1988). *Statistical power analysis for the behavioral sciences* (Vol. 2). Hillsdale, NJ: Lawrence Earlbaum Associates.
4. Computing Research Association. (2005). *Cyber-infrastructure for Education and Learning for the Future: A vision and research agenda*. Washington, D.C.
5. Custard, M., & Sumner, T. (2005). Using Machine Learning to Support Quality Judgments. D-Lib Magazine, 11, 11.
6. Fitzgerald, M.A., Lovin, V., & Branch, R.M. (2003). A Gateway to Educational Materials: An Evaluation of an Online Resource for Teachers and an Exploration of User Behaviors. *Journal of Technology and Teacher Education*, 11, 1, 21-51.
7. Glaser, B. G. and A. L. Strauss (1967). *The discovery of grounded theory: Strategies for qualitative research.* Chicago, Aldine Publishing Company.
8. Giersch, S., Leary, H., Palmer, B., Recker, M. (2008)a. Supporting Meaningful Learning with Online Resources: Developing a Review Process. In *Proceedings of the Annual Open Education Conference*, Logan, Utah, September 24-26.
9. Giersch, S. Leary, H., Palmer, B., & M., Recker (2008)b. Developing a Review Process for Online Resources. In *Proceedings of the Joint Conference on Digital Libraries,* Pittsburgh, PA, June 15-20.
10. Hanson, K. & Carlson, B. (2005). *Effective Access Report.* Education Development Center Instructional Architect Quality Guide, latest version: http://digitalcommons.usu.edu/itls_research/5/
11. Kastens, K., DeFelice, B., Devaul, H., DiLeonardo, C., Ginger, K., Larsen, S., et al. (2005). Questions & Challenges Arising in Building the Collection of a Digital Library for Education: Lessons from Five Years of DLESE. *D-Lib Magazine*, 11, 11.

12. Knox, D., et al (1999). The Peer Review Process of Teaching Materials: Report of the ITiCSE'99 Working Group on Validation of the quality of teaching materials. *Annual Joint Conference Integrating Technology into Computer Science Education*, Cracow, Poland.
13. Lamantia, J. (2003). *Analyzing Card Sort Results with a Spreadsheet Template*. Retrieved January 18, 2008, from Boxes and Arrows Web site: http://www.boxesandarrows.com/view/ analyzing_card_sort_ results_with_a_spreadsheet_template
14. Leary, H., Giersch, S., Walker, A., Recker, M. (2009). Developing a Review Rubric for Learning Resources in Digital Libraries. *ITLS Faculty Publications*. Paper 17. http://digitalcommons. usu.edu/itls_facpub/17
15. Liu, K. & Ward, V. (2007). Access Excellence @ the National Health Museum. *D-Lib Magazine*, 13, 11/12.
16. Maurer, D. & Warfel, T. (2004). *Card Sorting: a definitive guide*. Retrieved January 18, 2008, from Boxes and Arrows Web site: http://www.boxesandarrows.com/view/card_sorting_a_defini tive_guide
17. McMartin, F. (2004). MERLOT: A Model for User Involvement in Digital Library Design and Implementation. *Journal of Digital Information*, 5, 3.
18. Muramatsu, B. & Agogino, A. (1999). The National Engineering Education Delivery System: A Digital Library for Engineering Education. *D-Lib Magazine*, 4, 5.
19. Nesbit, J., Belfer, K., Leacock, T. (2003). *Learning Object Review Instrument (LORI 1.5)*, User Manual. Retrieved January 15, 2008.
20. Ohio Resources Center (ORC): http://ohiorc.org
21. Pea, R., with Christine L. Borgman (Chair), Hal Abelson, Lee Dirks, Roberta Johnson, Kenneth R. Koedinger, Marcia C. Linn, Clifford A. Lynch, Diana G. Oblinger, Katie Salen, Marshall S. Smith, Alex Szalay (2008). *Fostering learning in the networked world—the cyberlearning opportunity and challenge: A 21st century agenda for the National Science Foundation (Report of the NSF Task Force on Cyberlearning)*. Arlington, VA: National Science Foundation.
22. Recker, M., Leary, H., Walker, A., Diekema, A. R., Wetzler, P., Sumner, T., Martin, J. (2011, April). Modeling Teacher Ratings of Online Resources: A Human-Machine Approach to Quality.

Paper *presentation at the American Educational Research Association*, New Orleans.

23. Recker, M., Walker, A., Giersch, S., Mao, X., Halioris, S., Palmer, B., Johnson, D., Leary, H., Robertshaw, M.B. (2007). A study of teachers' use of online learning resources to design classroom activities. *New Review of Hypermedia and Multimedia*, 13, 2, 117-134.
24. Recker, M. (2006). Perspectives on teachers as digital library users: Consumers, contributors, and designers. *D-Lib Magazine, 12*, 9.
25. Recker, M., Walker, A,. & Lawless, K. (2003). What do you recommend? Implementation and analyses of collaborative information filtering of web resources for education. *Instructional Science*, 31, 299–316.
26. Robertshaw, M. B., Leary, H., Walker, A., Bloxham, K., Recker, M. (2009). Reciprocal mentoring "in the wild": A retrospective, comparative case study of ICT teacher professional development. In E. Stacey (Ed.) *Effective Blended Learning Practices: Evidence-Based Perspectives in ICT-Facilitated Education*, Melbourne: IGI Global Press.
27. Sumner, T., Khoo, M., Recker, M., & Marlino, M. (2003). Understanding educator perceptions of "quality" in digital libraries. In *Proceedings of the Joint Conference on Digital Libraries*, Houston, Texas, May 27-31.
28. Walker, A., Recker, M., Lawless, K., & Wiley, D. (2004). Collaborative information filtering: A review and an educational application. *International Journal of Artifical Intelligence and Education*, 14, 1-26.

5

CHAPTER

CONVERGENCE AND DIVERGENCE AMONG DIGITAL LIBRARIES AND THE PUBLISHING INDUSTRY

Terence K. Huwe

Institute for Research on Labor and Employment Library, University of California, Berkeley United States of America

INTRODUCTION

The Internet's dynamic impact on society, industries and individuals has been studied intensively across a broad number of academic disciplines. Digital media are spurring both creativity and dislocation in every field of study, as well as the workplace. These forces have also triggered sweeping changes in how traditional players in the creation of knowledge and scholarship operate and interact with each other. "Digital convergence," as it is widely known, invites not only creativity and enterprise, but also new and energetic competition in nearly every line of work (Yoffie, 1996).

The evolving roles of the traditional publisher and the research library in the United States are particularly illuminating as indicators of the ferment that the digital era is producing. These two groups have long enjoyed ties of mutual benefit, but now face radical forces of change, and find themselves in competition with each other—as digital publishers. Indeed, it is now possible for each agent to assume the characteristics of the other: librarians can act as digital publishers, and publishers can add a few percentage points of total book sales—and book sales, although they have been shrinking since 2004, nonetheless generated more than 25 billion dollars in 2009 (Datamonitor, 2003, 2010). Moreover, 2010 appears to be the year that acceptance of e-book devices and electronic text reading will accelerate in popularity. By the end of 2010, 10.3 million people are forecast to own an e-reader, and will buy as many as 100 million e-books (Richtel & Miller, 2010). With such rapid trends underway, the influential Book Industry Study Group has devoted considerable energy and resources to study the whole of the industry, with a close focus on electronic products (Healy, 2008). Publishers have also have released content in aggregated and topical "libraries" such as Wiley Custom Select, which offers a course- reader solution to professors (WileyPLUS, 2010). In short, the business environment, though fraught with challenges, continues to be functional, and publisher strategies have been evolving at a faster pace (Brown & Boulderstone, 2008).

Although new ideas are being actively studied, staff morale continues to suffer. Attention to the publisher "death watch" and a "rallying the troops" rhetorical style continue to dominate the tenor of publishing's professional dialogue. In 2010, the Book Industry Study Group conducted an important survey of publishing staff, to gauge their perspectives and sentiments regarding new media and e-books. When asked when they expected to see fundamental change in their own functional work area, 31 percent of respondents replied that it had "already happened", and 45 percent said it was "happening now" (Schatzkin, 2010). At the same time, enterprising digital publishers such as O'Reilly Media have countered with comprehensive business models that draw on the full array of social media, creating interactive user experiences that go beyond the book format (Hane, 2010). By 2008, the full force of experimentation and pursuit of innovation had gripped the publishing industry, creating a sense of making up for lost time.

DIGITAL LIBRARIANS: PRESERVING THE PAST, LOOKING FORWARD

Throughout the same eras of upheaval and change, librarians faced similarly daunting challenges. Clarion calls heralding the imminent demise of a proud profession crowded the professional literature of every specialized sector of librarianship for decades (White, 1989, Lowry, 2001). In response, various innovators have explored a mulitiplicity of strategies. Nancy Lemon (1996) argues that rethinking traditional roles enables librarians to climb organizational "value chains;" James Matarazzo and Toby Pearlstein (2010) review the plight of news media libraries and see a history of cyclical renewal, in which existing jobs give way to new opportunities in new locations. Among corporate libraries, a wide variety of bare- knuckle strategies have been put forward, urging librarians to find a competitive edge by offering services that can be demonstrated to boost profits (Chandler & Carroll, 2002). In the public libary sphere, despite straitened civic budgets and the shock of new media, libraries have achieved a degree of success in staying relevant. Circulation, library card membership and on-site use of services are at an all-time high, countering the idea that community libraries are outdated during the digital era. On the contrary, they are more popular than ever (Nolte, 2010). Therefore even as pessimism about the future became a standard feature in the professional literature, a parallel stream of daring and innovative thinking has run concurrently among a wide spectrum of library specialists.

The resulting intellectual ferment has produced new paradigms and new energy for rethinking library goals in light of the emergence of digital technology. Rather than wait to see what would happen, librarians repeatedly took initiative with new media. They staked an early claim on the crucial issue of intellectual property and copyright, working both with and against publishers as needed to preserve the library's public service mission. As part of this process, librarians became more deeply aware of the publishing industry's travails with digital media. This awareness led to greater strategic knowledge about the marketplace. (Fister, 2001; Katz, 2010; Hane, 2010; Howard, 2008).

Spanning "online" eras

Librarians were shaken by the rapid emergence of Web-based information resources, because they had been important players in the previous "online" economy that was dominated by firms like LEXIS/NEXIS and Dialog Information Services. This first "online" era gave librarians the opportunity to cast themselves as experts who conducted mediated searches on behalf of users such as attorneys, scientists and business leaders. As the initial "online" era faded and the Internet exploded into growth, librarians joined the first wave of Internet e-mail "conversationalists", Web content users, Web site producers, and aggregators of high quality information. They also grasped the importance of creating stable and robust Web portals, which organized and "branded" library-hosted aggregations of databases; such activist strategies stretched scarce budgets while formalizing a Web-based library presence (Howard, 2009). Similarly, the economic upheaval caused by the skyrocketing price of scholarly journals generated energetic responses on the part of research libraries, including formation of the Scholarly Publishing and Academic Resources Coalition (SPARC), as well as active congressional lobbying to protect copyright and "fair use" principles from a runaway marketplace. The same upheaval also sparked a sustained outreach to faculty, in search of much stronger partnerships with the academy's principle content creators. In general, digital librarians have been agents for innovation over the past 15 years, and have participated in the sweeping process of rethinking scholarly communications (SPARC, 2010; Berkeley Research Impact Initiative, 2009).

This lengthy process of trial and error has transformed digital librarians' self-perception. Rather than viewing digital media as a force of dislocation that would bring ruin to the profession, digital librarians have instead embraced it. Likewise, constant downward pressure on budgets forced innovative survival plans to the forefront, emphasizing new technologies (Howard, 2009, McKenzie, 2009).

Metadata as enhanced competency

Perhaps the most significant advance on the part of digital librarians has been the profession's embrace of structured metadata and taxonomy as a reinvigorated core competency. Even as they faced the twin challenges of integrating new technologies and shrinking budgets, digital librarians participated in the formation of international metadata standards such as the Dublin Core (Dublin Core Metadata Intiative, 2010). Libraries became key institutional members of new groups such as the Coalition for Networked Information and the National Digital Library Federation. Academic librarians experimented with diverse forms of preservation, including digital repositories, e-journals, and increasingly, full-scale publishing initiatives that originate within the library (Furlough, 2009).

Many metadata platforms and languages have been explored, but

Extensible Markup Language (XML) has become a dominant tool for managing digital assets (OASIS, 2010). XML is a "meta-language of languages," enabling developers to create taxonomies and metadata schema that are customizable, portable and attached to the "digital objects" they describe. The emergence of national and international standards for metadata, the ascendance of digital objects which are transferrable among collections, and a strong focus on systems interoperability have elevated the library profession to a technical status considerably higher than it enjoyed before the advent of the digital era (Johnson, 2010). Although digital librarians are not the only XML developers—many computing firms and other academics are heavily involved—they have left an imprint on the development stream of XML.

XML-based systems have become accepted across many industries beyond the academy, and now play a crucial role in information management and data warehousing by large corporate firms. XML-based information architectures are essential tools for managing text, images and other artifacts, creating new workflows that are based on the principle of "one text, multiple outputs". This kind of workflow would transfer very effectively to the publishing world, yet the publishing industry has been slower to adopt XML workflows at a universal level. The opportunity to manage digital assets using XML is an important leap forward that publishers have not yet fully embraced, and forward-looking commentators have recognized this shortfall (Ganesan, 2010; Young & Madans, 2009).

From online repository to publishing platform

The turmoil of the Internet's early years also produced solid initiatives to understand the digital library as a publisher in its own right. Thomas (2006) provides an early and comprehensive road map that offers large research libraries a template for launching high quality publishing services. She also describes the early emergence of online repositories as a new form of curated information resources. Online repositories grew quickly throughout the early years of the twenty-first century, taking much inspiration from the digital archiving efforts of the computer science and engineering fields (Furlough, 2010). Library-managed repositories also grew up with a strong bias for interoperability, which has led to more dialogue about the need for large-scale "federations" of digital libraries, even at the international level (Van de Sompel, 2006, et al).

As these repositories became more accepted, use skyrocted. The University of California's eScholarship repository experienced more than one million downloads in its first two years of operation. "Post prints" and research reports are now also collected in online repositories, and are frequently overseen by digital librarians, or at least coordinated by them. In late 2009, eScholarship recast itself as a full-scale "publishing platform," which drew strength from the reputation of its source—the University of California system. Although this is not the only example of a library-sponsored publishing service with university imprimatur, it is certainly one of the most high profile experiments (SPARC, 2006).

The evolution of library-based online repositories and their current transformation into full- scale publishing platforms is a prime example of how contributors can expand their role and move into new "zones of progressive release" (Abbott, 1988, 1991). Digital librarians' publishing solutions are guided by two objectives: first, to empower authors, and second, to create robust, collaborative blocs of institutions that share expertise and improve access. These are new "treatments" that address the changing needs of academic publishing (Van de Sompel, et al, 2006). Publishers are rushing to reinvigorate their relationships with authors, but to some degree they are playing a game of catch-up.

CONVERGENCE AND DIVERGENCE: STRATEGIES AND EXAMPLES

Recent events in these two fields reveal several interesting trends, not only in strategic planning but also in the underlying thinking of leading commentators. Although there is considerable ferment and thus many trends underway, there are four areas where publishers and digital librarians face similar challenges. The strategic decisions that each group makes in the coming years will have significant impact on their long-term futures.

First, the hurdles to publishing a well-packaged text artifact are dropping rapidly. The result has been that authors, both in the academic and popular literature spheres, now have the option of creating their own digital artifacts without publishers' assistance. Many already manage online presences and engage in dialogue with readers. In response publishers and digital librarians are testing strategies that reinforce their own roles.

Second, the disruptive nature of digital media has forced publishers and digital librarians to evaluate their own native skill sets in a new light. The library profession's core competencies—collection development, information counselling, interpretation and preservation—involve in-depth analysis of information resources. Yet librarians already format, revise, copy-edit and even print bibliographies, scholarly e-journals, festschrift, and full-scale books in many cases. As these content-intensive roles become more important, research libraries have responded by creating senior management positions with titles such as director of digital scholarly publishing, director of digital publishing, or director of built content. These positions carry responsibility for assessing new opportunities to publish, as well as assisting research faculty in doing so themselves.

Publishers also have the opportunity to review what they do well, re-evaluating existing functions to include new services such content aggregation, developing much-enlarged Web presences, taking on custodial roles and offering services for long-term preservation of their own built content. The technological hurdles to adding new services of this nature are just as low for publishers as they are for librarians who are involved in digital publishing. Both groups are limited chiefly by imagination, by perceived commitments to doing "business as usual," and by the high cost of retraining their already-well-trained work forces to take on new tasks in addition to their existing workflows.

Third, both publishers and digital librarians are looking beyond the boundaries of their own fields of expertise for new ideas and strategies. For example, new social media have carried a heavy impact on the news industry, and both librarians and publishers have studied newspapers as a cautionary tale. News media were one of the first zones where readers began "talking back" to the press and contributing substantive new content (as well as trivial or satirical interactions). The Blogosphere has also attracted interest, although it increasingly produces voluminous and cyclical "blooms" of creative work that are followed by inactivity and the formation of "dead zones." This pattern suggests that new media are only as vibrant as the minds that are driving them forward (The Economist, 2010). However, over time the Blogosphere has become firmly established as an effective platform for commentary, news, cultural critique and debate. The Blogosphere's experience implies that longer-term value takes time to emerge as a new technology matures.

Fourth, both groups have come to realize that as digital convergence has accelerated, bold action is required. As a consequence, the pace of intellectual thought that is devoted to innovation has also accelerated. There is also greater acceptance that bold actions may succeed or fail, yet they must be attempted to gain new knowledge and expertise. Both groups show evidence of complex responses to the necessity of taking bold action, because both groups believe that they must protect legacy print programs, whether as book sales or print collections, even as they step into new digital futures. This is a difficult balancing act, because reader and user community loyalty may be challenged as risks are taken.

These four trends illustrate the turmoil, and indeed, the excitement of the times for both publishers and librarians. They also illustrate how each group is operating on the knowledge creation continuum, and how they might choose convergent or divergent strategies when compared to each other.

Convergent strategies

With respect to convergent strategies, publishers and digital librarians are aware that their core competencies must now include a robust and dynamic "conversation" with their readers, collaborators and user communities. Although this may seem obvious during the Internet era, it nonetheless symbolizes a major challenge for established players on the knowledge creation continuum. New ideas and new technologies, commonly known as "Web 2.0", currently emphasize ubiquitous interaction in a wide variety of locations and via a long list of tools, ranging from desktop computers to "smartphones."

Among publishers, this interactive paradigm and the tools it has spawned have caused a seismic shift in thinking. A book's usefulness is no longer limited to the experience of reading it; it now has a lifespan that can take many forms, involve communities in addition to solo readers, and last for years (Norrington, 2010). In response publishers have launched serious attempts to enter the Web 2.0 sphere. Likewise, digital librarians have seen their print-based mission expand exponentially to include not only the finished works of scholarship and literature, but also the artifacts created by the overall process of creating scholarship, from start to finish. Library responses also include innovative combinations of digital collections and community-building features, such as allowing commentary, running newsfeeds, and adding Wikis to facilitate dialogue.

Publishers continue to focus on sales and profits, and they are experimenting with social media to increase revenue. Two strategies dominate the landscape: narrowcasting and community-subscriber services (Kist, 2008). Narrowcasting refers to the strategy of discovering discrete markets or user communities who share strong interest in very specific literature, and then offering them targeted products that are based on market analysis (Shaver and Shaver, 2009). This approach is greatly assisted by "viral marketing," a common term in the Internet era, which describes how news of events or products can travel very quickly, even circumnavigating the globe in a matter of hours in some instances. Narrowcasting would be paired with general marketing strategies, just as print sales would be complemented by e-book sales, still a small (but growing) revenue stream.

Community-subscriber based strategies also make explicit publishers' new role as their own distribution outlets, joining bookstores, libraries and online firms such as Amazon in direct consumer outreach. Direct outreach is another example of treatment substitution, as it establishes a new zone of service on the knowledge creation continuum for publishers, shifting them further into the zone of distribution and perhaps even preservation.

Recent developments in textbook publishing provide evidence of innovative approaches by publishers and digital librarians, who are exploring classroom teaching aids. It is now possible to print textbooks or sections of them on demand, use e-readers to read them, or purchase printed readers, and the entire idea of how textbooks support teaching is rapidly evolving. Publishers are now perfecting "portal" style learning zones on the Web, which are based on textbooks but include many added features, including unbundled chapters, added teaching aids and accompanying training modules. Interestingly, 55 percent of students still prefer to buy the textbook in print as part of their study plan (Vance, 2010; WileyPLUS, 2010). Meanwhile, digital librarians are also involved in teaching portals, and they have been perfecting e-reserve systems and new formats for class e-readers (bSpace, 2010).

Finally, publishers are beginning to show interest in managing their backlists more along the lines of a repository or collection of resources.

However, whether publishers will take up archiving and preserving content remains uncertain. Once again, networked information technologies have lowered the hurdles to creating online archives. But in practical terms, taking on an archival role would require publishing staff to learn new skills, or recruit new talent to join the firm. Early evidence suggests that publishers have not fully embraced the link between the process of acquiring, editing and selling books and the long-term value of archiving the material (Schatzkin, 2010). This further suggests that publishers continue to regard themselves as facilitators of the early stages of a book's lifespan, but not as the custodians of its entire lifespan.

Divergent strategies

There is a large common ground of shared strategies among publishers and digital librarians, derived primarily from the interactive nature of social media, and how it may be adapted for research or to enhance popular literature. However, the points of divergence between the two groups are pronounced. Divergent strategies flow directly from the history of each group.

Digital librarians are working very hard to preserve and advance the role of managed knowledge resources, branded by the library, as part of the teaching process. In addition to experimentation with e-reserves and e-readers, they are now staking a large claim on the full-service teaching Web "portals." Open-source instructional portals now include a variety of added functions, including the ability at attach related files, images and simulations (bSpace, 2010). They also operate as eportfolios for students that follow them throughout their academic careers, and as information management tools for the faculty. Academic librarians perceive important new roles for information services in these learning spaces.

Academic librarians are also brainstorming about ways to enhance "built" content that is created by the faculty—a key zone of knowledge creation where libraries may assume the role of digital publishers. The opportunities are vast, as pre-publication content creation encompasses the supporting information, texts and data sets that lead to finished work. Strategies to preserve this knowledge base are rapidly taking shape, and they are evidence of innovative thinking about librarian core competencies (Abels et al, 2003).

What is more, the library profession's original core competencies—

particularly collection development and classification—gain new relevance and importance as digital publishing moves to the forefront. Metadata schemes are vital tools for managing vast amounts of digital assets. The prevailing scheme, the Metadata Enhanced Technical Standard (METS), has seen heavy involvement by digital librarians; it ensures that metadata are portable and stay attached to a digital artifact, allowing the metadata and the object they describe to migrate over time (Library of Congress, 2010). In contrast, publishers are in the early stages of harnessing XML to manage content more flexibly (Ganesan, 2010).

The most significant divergent characteristic between publishers and digital librarians has been librarians' willingness to enter into collaborative alliances, launching aggressive outreach to faculty authors, building political lobbying groups, and forming consortia that negotiate for better prices. They also have become software developers at their host universities, emphasizing open-source computing (Van de Sompel,2006). Ming-xing Huang (2010) envisions even more broad alliances, called "Digital Library Alliances"—which would enable academic libraries and their partners to enhance digitization initiatives and search capabilities in ways that mimic large commercial firms such as Google. Digital librarians have also sought explicit partnerships with publishers themselves, when a shared goal could be seen. For example, the California Digital Library (CDL) entered into an early agreement with the Berkeley Electronic Press (BePress)—a full service journal publishing solution (BePress, 2010, eScholarship, 2010).

Conversely, publishers' efforts to form large scale collaborations have taken more measured steps. With respect to libraries, collaborative efforts most often take the shape of advisory committees, which meet with editors and publishers once or twice per year. This has been a useful process, contributing to several significant joint efforts, such as the Wiley Online Library (Wiley Online Library, 2010). Wiley's new "learning space" includes extensive links to library services for training and other assistance in using the aggregation of content.

Digital librarians could afford to choose a collaborative stance, because they have been able to draw on strong relationships with their user communities. Historically, library patrons would visit a library in person, creating opportunities for a direct, personal relationship with well-trained professional staff. Armed with very good metrics on what library patrons actually need and how they prefer to gain access to resources, digital librarians are exploring how to create "user experiences" that reinforce a bond between the library and the user.

Just as important, the library profession conceptualized the digital library as a matrix of content, services and human interactions from the earliest planning process. In essence they have argued that a digital library is far more than a content platform; it is an entire community (Lyman, 1996). For example, facilitating and teaching how to use of information services, whether print or online, is a measurable core competency for librarians. They have argued that technology should enhance human connectivity, rather than replace it, and they have backed up this claim with solid e-metrics showing how people are using libraries.

This assertion of the digital library as community is not nearly as evident in the professional literature of publishing. However, publishers are beginning to reach similar realizations about digital media, and are examining different approaches. As online publishing creates new synergies between print and electronic artifacts, the book gains a broader venue for discovery: the Web itself. Publishers continue to evaluate reader responses to books and related Web sites; and while print books will continue to exist, new zones are opening up for books to grow via Web sites, as e-books, and as subjects of reader forums. If indeed these trends follow their current course, the digital or Web version of a given book may eventually become the "master copy of record." (Kist, 2008).

COLLABORATE, COMPETE, OR BOTH?

Digital librarians possess all of the tools and expertise needed to compete directly with publishers. As universities accelerate their plans to create open-source journals and lend their imprimatur to them, digital librarians may take key leadership roles in managing the new archives and repositories, perhaps even the most central role of content owner. If so, they will be building upon existing relationships with the faculty; the two groups work in close proximity and in many cases share research interests.

Digital libraries also share the advantage of their host institution's imprimatur. This prestige enables them to expand their initiatives and to gain institutional support in doing so. This trend is already underway; if it accelerates, the trend would carry multiple benefits for digital librarians (Hahn, 2008, 2008a). First, it would cement and formalize their new role as a digital publisher within the academy. Second, the new status conferred by the role of digital publisher will provide digital librarians with a fresh opportunity to argue the value points of their longstanding core competencies. Third, recognition of editorial work as a library skill will advance the professional status of digital librarians in the eyes of the research faculty and within the university administration.

All of these enhancements to the status of digital librarians within the academy are excellent examples of "treatment substitution" as described by Abbott (1988). Not only is the publisher role being offered by a competing group—digital librarians—it is also being used as a springboard to advance the status of the library profession as a whole.

Impetus for collaboration

Even though digital libraries' enhanced imprimatur strengthens their chances of becoming effective digital publishers, evidence indicates that they remain quite receptive to collaboration, often proposing complimentary services in dialogue with publishers (Hahn, 2008, 2008a). This suggests that digital librarians could become partners in more ambitious alliances with publishers, structuring them to preserve revenue streams while advancing user access. Such a strategy would serve as a means for using digital media to find solutions to the longstanding problems of the former print-only era, with advantages for both parties. At present, digital librarians are exploring strategies to manage the entire lifespan of knowledge creation and the materials that lead to a finished work. These include data sets, simulations, and non-text artifacts of every sort. It is also increasingly common for research libraries to manage their own data functions and image collections, and work closely with other campus organizations that are involved in similar work (Whalley, 2010). The role of data manager is being assumed by many innovators, not only at colleges and universities but also at the Library of Congress and other national-level collections.

The digital librarian-as-data manager could be a very powerful ally for publishers who seek to transform their books into extended "dialogues" with readers, including related data resources and

strategies, the arsenal of strategy at their disposal is significant.

CONCLUSIONS

Publishers and libraries have enjoyed strong links over the years, marked by moments of collaboration as well as competition. The urge to compete has accelerated due to the impact of digital media, and the increased ease of launching digital publishing initiatives. Moreover, both groups have skills within each other's core functions. Publishers are exploring how to manage content over time, and to find new value in their backlists. Digital librarians have overseen well-established publishing programs, often linked to special collections, and this provides them with skills for launching digital publishing programs. Against this backdrop, both groups are evaluating whether their core skills should expand to include roles that encompass the full lifespan of creative and scholarly works.

This process has been driven by technological evolution, and the forces of digital convergence. The processes that govern competition, collaboration or a combination of both have been well-studied by sociologists such as Andrew Abbott (1988, 1991), whose theories suggest that ongoing competition between these two knowledge-creating groups is quite likely. Even though the process of publishing is distinct from the practice of librarianship, the workflows and intellectual activities of both of groups are closely connected. Taken together, they encompass a large zone of influence on the knowledge creation continuum.

Digital media make explicit the linkages between the processes of publishing, and the library-centric processes of information counselling, interpretation and preservation. These linkages are increasingly apparent to both groups, inviting serious study of their future options, as commercial, educational and entertainment markets continue to evolve. The rewards for success in creating expanded information management roles are also apparent, both from profit perspectives and as a means of increasing prestige. The forum of competition has increased beyond the well-known debate about journal-pricing and open source publishing programs, and now includes opportunities for attracting user and reader "attention" with Web 2.0 technologies. All of these factors suggest that publishers—both trade and professional—

will find themselves looking at the library field for fresh ideas, and that the reverse will occur among librarians.

With the reduction of the technological barriers, the remaining obstacles are fundamentally organizational or cultural. The temptation to perpetuate known ways of managing workflows may obscure new opportunities for either group to make bold moves and take on new roles. Likewise the personnel expense of adding new functions—such as repository management for publishers, or greatly-increased editorial roles for digital librarians—is another hurdle. If organizational or cultural factors hinder strategic thinking, it is possible that collaboration between the two groups may increase as they struggle to innovate.

Even though Abbott's theory of treatment substitution augurs long-term competition in many forms, the outcomes are far from certain. Strategic planning among both publishers and digital librarians is crucial for creating advantage and reformulating their professional visions for the future. There are three areas to monitor as early indicators of how competition will play out. These include the direction of digital textbook and e-book design and function, since technical innovations may originate from either group; the advance of interactive repositories that increase the value of original creative works; and the formation of new workflow strategies to repurpose existing skills and add new functions. The strategic choices of each group will carry wide impact on the design of digital libraries, and therefore the processes of convergence and divergence among the two groups are worthy indicators for study by all stakeholders in digital library development.

REFERENCES

1. Abbott, A. (1988). *The System of Professions.* University of Chicago Press, 978--2260-00695, Chicago Abbott, A. (1991). The order of professionalization: An empirical analysis. *Work and Occupations,* Vol. 18, No. 4 (November 1991) 355-384, 0730-8884
2. Abels, E; Jones, R; Latham, J; Magnoni, D.; Marshall, J (2003). *Competencies for InformationProfessionals of the 21st Century.* SLA, Inc, retrieved from http://www.sla.org/competencies1997 on July 21, 2010

3. Banks, M. (2006). Towards a continuum of scholarship: The eventual collapse of the distinction between gray and non-grey literature. *Publishing Research Quarterly*, Vol. 22, No. 1 (March 2006) 1-11, 1053-8801

4. Berkeley Research Impact Initiative (2010). Retrieved on August 1, 2010 from http://www.lib.berkeley.edu/scholarlycommunication/
5. BePress (2010). The Berkeley Electronic Press: see http://www.bepress.com
6. Bosman, J. (2010). Biggest U.S. book chain weighs sale. *The New York Times*, August 4, 2010, p. B4, 0362-4331
7. Bosman, J. (2010a). Quick change in strategy for a bookseller. *The New York Times*, August 12, 2010, p. B1, 0362-4331
8. Brown, D. & Boulderstone, R. (2008). *The Impact of Electronic Publishing: The future for publishers and librarians*, Saur, 978-3-598-11515-8, Munich
9. bSpace (2010). Retrieved on September 2, 2010 from http://ets.berkeley.edu/bspace
10. Cader, M. (2008). Innovation is iteration: Thinking next to the box. *Publisher Research Quarterly*, Vol. 24, No. 4 (December 2008) 240-250, 1053-8801
11. Chandler, Y., & Carroll, C. (2002). Libraries and librarians: The key to growth and survival? The relationship between corporate productivity and information services. *INSPEL*, Vol. 36, No. 4 (December, 2002) 223-254, 0019-0217
12. Datamonitor (2003). *Publishing in the United States.* Retrieved from http://www.datamonitor.com
13. Datamonitor (2010). *Publishing in the United States.* Retrieved from http://www.datamonitor.com
14. Dublin Core Metadata Initiative (2010). See http://www.dublincore.org, retrieved on July 26, 2010
15. The Economist (2010). The empire slips away. *The Economist*, Vol. 395, No. 8688 (June 24-July 2, 2010) 62, 0013-0613
16. Ellison, J and Eatman, T. (2008). *Scholarship in Public: Knowledge Creation and Tenure Policy in the Engaged University.* Syracuse, NY, Imagining America, retrieved on August 3, 2010 from http://www.imaginingamerica.org/TTI/TTI_FINAL.pdf
17. Epstein, J. (2008). The end of the Gutenberg era. *Library Trends*, Vol. 57, No. 1 (Summer 2008) 8-16, 0024-2594

18. Epstein, J. (2010). Publishing: The revolutionary future. *New York Review of Books,* Vol. 57,No. 4 (March 11, 2010) 1-2, 0028-7504

19. eScholarship (2010). See http://escholarship.org/uc/ucias, retrieved on July 28, 2010 Furlough, M. (2009). What we talk about when we talk about repositories. *Reference & User Services Quarterly,* Vol. 49, No. 1 (Fall 2009) 18-32, 1094-9054

20. Fink, E. (1999). The Getty Information Institute: A retrospective. *D-Lib Magazine,* Vol. 5, No.3, (March 1999), 1082-9873. Retrieved from http://www.dlib.org/dlib/ march99/fink/03fink.html on July 21, 2010

21. Fister, B. (2001). Trade publishing: A view from the front. *portal: Libraries and the Academy,* Vol. 1, No. 4 (2001) 509-523, 1530-7131

22. Ganesan, D. (2010). The best reason for re-engineering book publishing—the need for XML. *TeleRead* (2010). Retrieved from http://teleread.com/2010/01/20

23. Hahn, K. (2008). Publishing services: An emerging role for libraries. *Educause Review,* Vol. 43, No. 6 (November/December 2008), 1527-6619

24. Hahn, K. (2008a). *Research library publishing services: New options for university publishing.* Association of Research Libraries, Washington, DC. Retrieved from http://www.arl.org/bm~doc/research-library-publishing-services.pdf

25. Hane, P. (2010). Eye on O'Reilly, OpenCourseWare, Ebooks, Blio and More. *Information Today,* Vol 27, No. 5 (May 2010) 7-11, 8755-5286

26. Healy, M. (2008). Experimentation and innovation in U.S. publishing today: Findings from the Book Industry Study Group. *Publisher Research Quarterly,* Vol. 24, No. 4 (December 2008) 233-239, 1053-8801

27. Howard, J. (2008). Scholarly presses discuss how they're adapting to a brave new e-world. *Chronicle of Higher Education,* Vol. 54, No. 44 (July 11, 2008), 0009-5982

28. Howard, J. (2009). Libraries explore big ideas to overcome small budgets. *Chronicle of Higher Education,* November 22, 2009, retrieved on July 30 from http: //chronicle.com/article/Libraries-Explore-Big-Ideas/49227/, 0009-5982

29. Huang, M., Xing, C., & Zhang, y. (2010) Supply chain management model for digital libraries. *The Electronic Library,* Vol. 28, No. 1 (2010) 29-37, 0264-0473
30. Institute for the Future (1996). *Twenty-First Century Organizations: Reconciling Control and Empowerment.* Institute for the Future, Menlo Park, CA
31. Johnson, M (2010). *The Book is Overdue: How Librarians and Cybrarians Can Save Us All.* Harper, 978-0061431609, New York
32. Jovanovich, Peter. (2009). Publishing in hard times. *Publisher Research Quarterly,* Vol. 25, No. 2 (June 2009) 67-72, 1053-8801
33. Katz, R. (2010). Scholars, scholarship, and the scholarly enterprise in the digital age. *Educause Review,* Vol. 45, No. 2 (March/April 2010) 44-56. 1527-6619
34. Kist, J. (2008) *New Thinking for 21st-century Publishers: Emerging Patterns and Evolving Strategems.* Chandos, 978-1-84334 445-2, Oxford, UK
35. La Force, T. (2008). Publishing death watch. *The Book Bench: Loose Leafs from the New Yorker Book Department,* December 5, 2008. Retrieved on July 21, 2010 from http://www.newyorker.com/online/blogs/books/2008/12/publishing-titl.html
36. Lemon, N. (2006). Climbing the value chain: A case study in rethinking the corporate library. *ONLINE,* Vol. 50, No. 10 (November-December, 2006) 50-55, 0146-5422
37. Lessig, L. (2005). The people own ideas! *Technology Review* Vol. 108, No. 6 (Winter, 2005) 46-53, 0040-1692
38. Library of Congress (2010). Metadata Encoding and Transmission Standard: Primer. http://www.loc.gov/standards/mets/METSPrimerRevised.pdf
39. Lowry, C. (2001). "The more things change..." *portal: Libraries and the academy,* Vol. 1, No. 4 (2001) 7-9, 1530-7131
40. Lyman, P. (1996). What is a Digital Library? Technology, Intellectual Property and the Public Interest." *Daedalus, Journal of the American Academy of Arts and Sciences,* Vol. 125, No. 4 (Fall 1996) 1-33, 0011-5266
41. MacDonald, S & Uribe, L. (2008). Libraries in the converging worlds of open data, e- research, and Web 2.0. *Online,* Vol. 32, No. 2 (March 2008), PAGES, 0146-5422
42. McKenzie, J, Dekker, H., Ford, G., Hurley, B., de Lorenzo, D., Ogden, B., Phillips, M., Stanton, T., Xue, S. (2009). Digital Collection

Development Task Force: Final Report. Retrieved on July 28, 2010 from http://lib.berkeley.edu/Collections/pdfs/ digital_collection_development_plan.pdf

43. Matarazzo, J. & Pearlstein, T. (2010). Survival lessons for libraries: Staying afloat in turbulent times. *Searcher*, Vol. 18, No. 4 (May, 2010) 48-53, 1070-4795

44. Nash, R. (2010). Publishing 2020. *Publisher Research Quarterly*, Vol. 26, No. 2 (June 2010) 114-118, 1053-8801

45. *New York Times* Knowledge Network (2010). Retrieved from http://www.nytimes knowhow.com.index.php/epublishing-certificate-program on August 30, 2010

46. Nolte, C. Print is not dead. *San Francisco Chronicle*, August 15, 2010, p. A2. 1932-8672

47. Nowatka, E. (2008). Our digital future. *Publisher Research Quarterly*, Vol 24, No. 3 (September 2008) 124-128, 1053-8801

48. Norrington, A. (2010). Harnessing 'e' in Storyworlds: Engage, enhance, experience, entertain. *Publisher Research Quarterly*, Vol. 26, No. 2 (June 2010) 96-105, 1053-8801

49. OASIS: Organization for the Advancement of Structured Information Standards (2010). See http://www.oasis-open.org/who/

50. Overdorf, M & Barragree, A. (2001). The impending disruption of the publishing industry. *Publishing Research Quarterly,* Vol. 17, No. 3 (Fall 2001) 3-18, 1053-8801

51. Regazzi, J & Caliguiri, N. (2006). Publisher and author partnerships: A changing landscape. *Learned Publishing,* Vol. 19, No. 3 (July 2006), 183-192, 0953-1513

52. Richtel, M, & Miller, C.M. Of two minds about books. *The New York Times*, September 2, 2010, p. B1, 0362-4331

53. Rosemont College (2010). Graduate Publishing Program. Retrieved on August 30, 2010 from http://www.rosemont.edu/gps2 /graduate/academics/publishing/study.php

54. Shatzkin, M. (2010). BISG's Making Information Pay 2010: Selected Survey Results. Retrieved on July 27, 2010 from http://ww.bisg.org/ contentweb /papers mip2010

55. Shaver D. & Shaver, A.S. (2009). Books and digital technology: A new industry model. *Journal of Media Economics*, Vol. 16, No. 2 (November 2009) 71-86, 0899-7764

3. LIBRARY3.0

School library is a private realm where dominating relational form is between students of the same school. Whereas, public library is a public realm, most of the visitors are strangers to each other, and the library's community activities constitute it as a parochial realm. Both types have to change their scope of services and tasks as a result of digitizing and digital swift. The evolution is not only technology driven due to that of Web and e-Learning "point oh", but also market driven as the core skills include creativity, interdisciplinary thinking, problem solving, and the ability to collaborate with others cannot be learned easily from books, but rather through learning-by-doing and social interaction in today's knowledge base economy. Not to mention most users work individually or within pre-organized groups who do not usually make new connections with co-present, unacquainted users (Bilandzic and Foth, 2013) [13] , the attraction of traditional library learning arena is diminishing. Even with the adaption of Web3.0, electronic library can provides a new way for end users to collect online contents concerning comments, articles, Bolgs, photos, and videos in one place through parsing the linguistic expressions provided without the intentionally invoke specific application programs (Chu and Yang, 2012) [22] , it cannot fully convert a traditional library into an information center, a meeting place, and a technology hub. Library should embrace Web3.0 and go beyond. The facilities of library thus have to be changed as well as the role of librarian. It seems that the role of librarian has been weakened in the Library2.0 environment, they must be aware of the use of artificial intelligence management tools and semantic web in Library (Ahmed, 2015) [23] . The new role of librarian is collecting & providing information and offering instruction in research techniques and information use rather than by physical presence as somewhere people go (Saw and Todd, 2007) [24] . They are not only the prominent apomediaries guiding library users on how best to locate, access and use credible information in myriad formats from diverse sources, at the point of need (Kwanya et al., 2013) [25] , but also facilitators of connecting users. To convert a library into a personalizable, intelligent, sensitive and living institution with seamless engagement of library users, librarians and subject experts on a federated network of information pathways, Kwanya et al. (2013) [25] , posited five principles of Library3.0, namely the library is intelligent, organized, federated networked, apomediated, and personalized. The meaning and attributes of each principle are listed in Table 2 below.

Table 2. Summary of library3.0/3.1 principles, attributes, and applied context-awareness (adapted from kwanya et al., 2013 [25] ; noh, 2013 [26] ; siemens, 2004 [15] ; and hussain, 2012 [21]).

Library3.0 principles	Attributes of Library3.0	Applied context-awareness	Principles of Connectivism
The library is intelligent: Self-renewing, flexible, functional, integrated, efficient, resilient, autonomous and sensitive (adaptive)	Self-renewing, flexible, functional, integrated, efficient, resilient, autonomous and sensitive (adaptive) via selective intelligence (wisdom of the expert), ontology, smart (intuitive) searching in natural language, artificial intelligence	Space on/off: Controlling the space according to user access. Turn off the light & other available technology once the user entering the space. Lobby: Direct users to their desired space using location based technology	Learning may reside in non-human appliances.
The library is organized: Turn the unorganized web of information into a systematic and usable body of knowledge by exhaustively describing and linking every piece of data to enable ease of access	Turning the unorganized web of information into a systematic and usable body of knowledge by exhaustively describing and linking every piece of data to enable ease of access via semantic Web/Search		Ability to see connections between fields, ideas, and concepts is a core skill. Nurturing and maintaining connections is needed to facilitate continual learning
The library is a federated network of information pathways: Unifies disparate information channels, formats and environments to ensure availability, accessibility, search ability and usability of credible information	Supports expressiveness and interoperability to create synergies between hitherto disparate information resources and systems via network of data (meaning), all-visible, all-accessible Web, and Cloud computing		Learning is a process of connecting specialized nodes or information sources. Currency of knowledge is important.
The library is apomediated: Stand by the users and guide them to high quality information and services	Stand by the users and guide them to high quality information and services without overbearing on the users via social media, peers, & librarians.	Reference & Safety services: Riding on location based technology, if the user spends over certain duration at specific location looking for specific subjects, recommendations will be sent to him/her accordingly. If the user doesn't move over a certain time, alert will be sent to library personnel.	
The library is "my library": Personalization similar to what various service providers such as search engines and the media were offering their clients on various Internet platforms	Personalization of library services via the design, management and delivery of content based on known, observed and predictive information. The opportunity to interact with other library users and to form or join groups to enrich their learning experience	Access service: The system detects the user identity by reading the RFID card, & then pushes the recommendations according to his/her borrowing history.	Learning and knowledge rests in diversity of opinions. Capacity to know more is more critical than what is currently known

To develop an intelligent system which can search and provide the services that users require in their current situation by analyzing and identifying the available contextual information such as the user's current location, time, people and devices in the vicinity, and the user's behavior and inputted data, Context-Awareness must be applied (Noh, 2013) [26] . He posited the use of context awareness as a core concept of the next generation digital libraries. Context is divided into "user context" and "environment context". The former represents the wide range of information described by the user and describing the user while the latter is the contextual information relating to the user's current user's current state, such as location or time zone, which cannot be described within the user's context (Hartmann, 2010) [27] . In short, Context-aware computing provides service to meet the needs of the users by recognizing certain situations, such as the user's location or environment. The author expounded the applications into five proposals: Applied context-awareness access service, Space on/off according to the user's access, Applied context-awareness lobby, Applied context-awareness reference service, and Applied context-awareness safety service. However, Noh (2013) [26] realized that Context-awareness has not been fully implemented in library yet, and thus the performance required further evaluation in following studies. The Principles of Library3.0, Context Awareness, and Connectivism are summarized in Table 2. It has been shown that the principles of Libray3.0, Context-Awareness, and Connectivism are aligned with each other.

Although the total number of library visitors has been declining in recent years as mentioned before, research has shown that the number of children borrowing books from libraries is at its highest in history. There are children zone where children can read, talk, and play on the floor. Studies have also shown that students prefer actual books to virtual books. Libraries remain to be the source of information, culture and wisdom (Gaiman, 2013) [28] . Another group of frequent library users are the elderly who may not be familiar with Web searching. They are healthy, retired, prefer to read newspaper and enjoy the air-con in the summer time but seldom make friends with other users. Regret that the elderly is committing more crimes than those aged 14 to 19 in recent Japan. This is because some elderly would be willing to commit crimes just to find companions and care in prison (The Economist, Victims' justice, 2015) [29] . Elderly loneliness is a common problem in this aging society. Library is a place which can mitigate this problem particularly for those who do not stay in elderly home by providing some group activities other than reading materials. For instance, just the like children zone, libraries can set up themed corners such as the smart senior room to group individuals with different interests and interactions; it creates companion opportunity for them. The smart senior room is equipped with rehabilitation computer programs in which seniors will be able to select music, digital instructor, 3D animation, so that participants can go through the program at their own pace. At the same time, interactive games involving the multi-touch display for multiple players; different activities, and events can be organized for interaction and fun. As such, in order to change libraries for books/medias in industrial society to libraries for man/human in networked/ knowledge society, it should provide different spaces/zones to cater the need for individual, quite space for reading, noisy space for group meeting, learning, meeting, discussion, and performance for mixed or different age groups, partnership with creative alliance to organize different types of function, and integrate with citizen services to attract users.

Schulz (2014) [30] realized the above societal demand. He posited a new model for the "Next" public library, providing "Experience, Discover, Participate, and Create" through digitization and individualization: Experience through inspiration space, Discover through learning space, Participate through meeting space, and Create through performance space. Inspiration space should open up to the irrational, emotional and chaotic by communicating a diversion of aesthetic experiences through storytelling or other artistic expressions within all kinds of media, cultural forms and genres. Learning space should provide experience and exploration of the world and thus strengthen their competences and possibilities through free and easy access to information and knowledge, music, play and many other activities. Meeting space is an open arena for meeting other people with different interests and values and encounter opinions that challenge you through discussions and debate in non-committal, random encounters through lounge décor with newspapers and café atmosphere as well as more organized meetings. Performance space can provide user interaction with others for inspiration and creation of new creative expressions in the encounter with art and culture through interactive games and writing-, sound- and video-workshops. Library services can then be tailored for personal needs and circumstances as well as giving users the opportunity to interact with other library users and to form or join groups to enrich their library experience (Storey, 2004) [31] . This echoed the findings of Aabø and Audunson (2012) [32] who explored how the libraries function as meeting places for across cultural, ethnic, generational, and social lines in a complex multicultural and digitized society. They found that a high proportion of the users constitute the library as a place related to work and education. Variety of users making the library stands out as a conglomerate of the community. Some people working alone use the library as their work space while some people in between jobs use it for job- searching. Some people periodically use the public library as an inspiring place to work. It is an extension of the school, home or work place. Social capital can thus be built up and that the process of building social capital can be part of a community development process (Kilpatrick et al., 2003) [33] . In fact, all these aligned with the principle of "Personalization" in Library3.0 (Kwanya et al., 2013) [25] . Schulz (2014) [30] further illustrated the integration of citizen services such as issuing passport, driving license, and health security card with library can bring citizen in contact with library. Some cities like Oslo and Helsinki built concert hall and museum side by side with library so as to create a new icon in the city. In fact, a number of new public libraries all over the world had adapted the "Space" concept in one way or the others in their designs. To name a few: Vennesla Library in Norway is located in the city centre provides café, open meeting places, and a small scene, the building combines a library and house of culture. Black Diamond Library in Denmark includes a 600-seat auditorium, the Queen's Hall, used for concerts, literary events, theatrical performances and conferences. Surry Hills Library in Australia specializes in fashion and design titles, as well as having a significant lesbian, gay, bisexual and transgender collection. Nam June Paik Library in Korea aims to promote "non-linear and random access to information, and its production beyond the prescribed consumption." The user of information becomes the producer, and the static contents of the library turns dynamic through spontaneous expression and juxtaposition of ideas. The interior design of Biblioteca España Library in Colombia could "decontextualize the individual from the poverty that is experienced in the outside." The Maranello Library in Italy is intended to provide its visitors an introspective experience and visual connection with nature. The library of Muyinga in Burundi was built with the local know-how and traditions of Muyinga in mind. A "hallway porch" along a circulation space is often a part of the Burundian traditional housing as it provides a shelter from heavy rains and harsh sun.

4. LIBRARY DEVELOPMENT IN REALITY

A full implementation of Library3.0 is still remote. In fact, there is no consensus on the actual definition of Library3.0. Build on the Library3.0 generation evolution graphical representation by Noh (2012) [34] , this paper is trying to elaborate the technology available in the market for library in Figure 1, followed by a more detail explanations in Table 3.

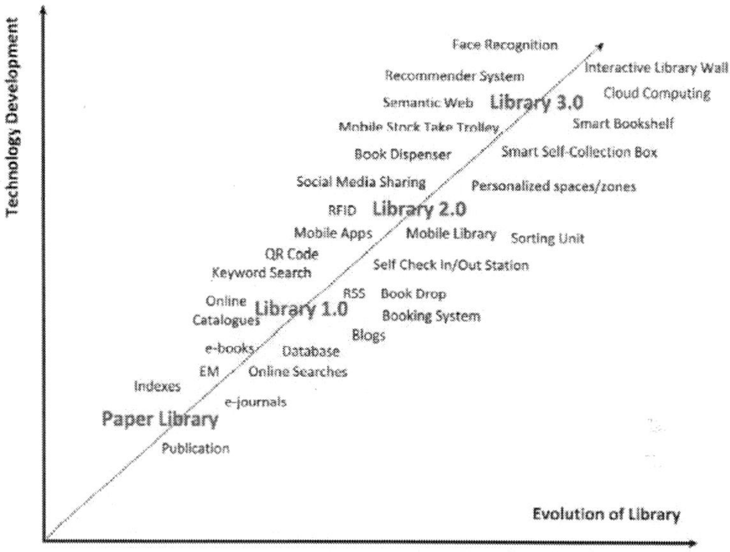

Figure 1. Evolution of library technology (adapted from Noh (2012) [34] with additional updates).

Table 3. Products for library3.0.

Intelligent & organized		
Product	**Features**	**Photo**
Library Automation	• Detects room occupancy • Automatic scene setting for different functions • Measures brightness & adjust lighting or curtain position • Detects temperature in room and adjusts to attain energy saving • Intercom system for communication with librarian • Integrated with room reservation & door access system	
RFID Applications	• RFID tags inside each book • Gate keeps track of RFID materials passing for security check • Counts the number of visitors	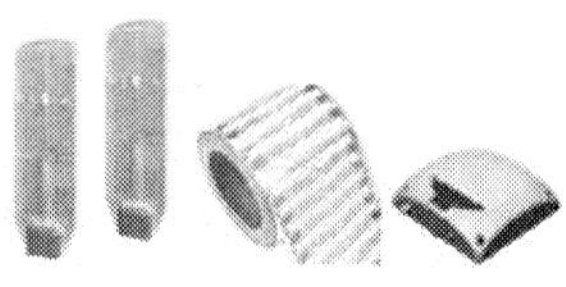

Smart Bookshelf	• Automatically detects on-shelve books • Updates system records for stock take • On-shelve books searching	
Smart Self-Collection Box	• Automatically distributes reserved books into dedicated smart shelf holes • Real time tracking and monitoring • Generates alerts in the case something has been misplaced	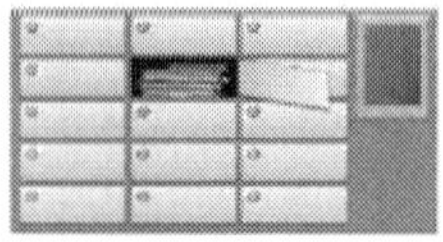
Mobile Stock Take Trolley	• Automatically detects books on the trolley for uploading to shelf • Provides stock functions with a handheld device	
Book Drop and Sorting Unit	• 24-hour book drop • Sorts books into their designated categories • Improves library efficiency and book location accuracy	

Continued

Booking System	• On-line booking of library facilities such as function rooms and multimedia equipment • Integrated with access control to authenticate users	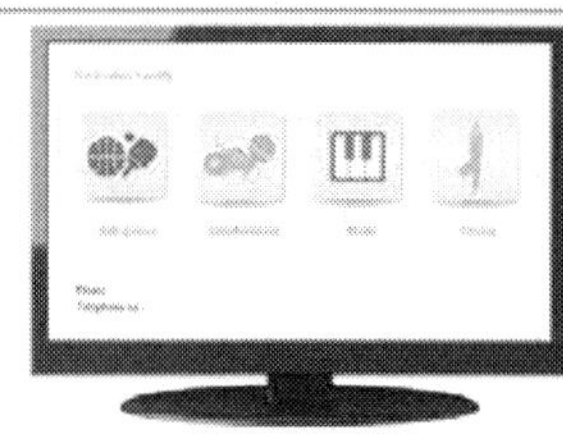
Self Check In/Out Station	• Encourages self-service • Hybrid system (HF/UHF/EM) to allow easy deployment and upgrades • Able to collect and analyzes user behavior	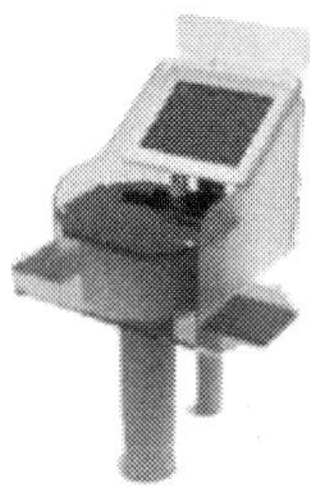

Federated Network of Information Pathways		
Product	**Features**	**Photo**
Book Dispenser	• Automatic detection of books on shelf • Updates the system records automatically • Self-service reader card registration • Self-service book check in/out • On shelve book browsing through touch screen monitor • Book reservation • Book recommendations	

Apomediated		
Product	**Features**	**Photo**
Intelligent Monitoring System	• Monitors abnormal user behavior • Integrated with security system to pop up problem location	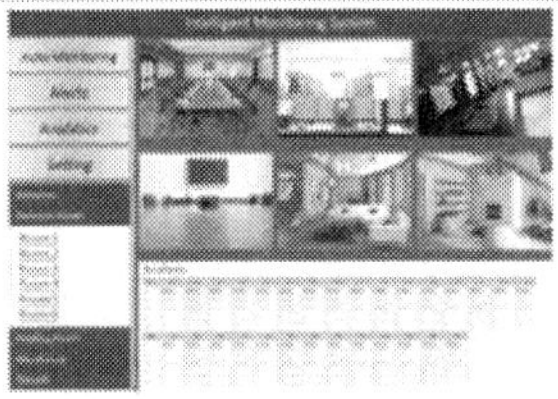
Real time location system	• Provides navigation function • Offers location-based promotions/services • Convenient searching • Reservation of library resources • Library map and guide • Event promotions • Enables social interactions • Integrated with recommender	

"My Library"		
Product	**Features**	**Photo**
Recommender System	• Recommends books based on user preferences • Identifies user's previous records. Content-based or/and collaborative filtering	
Face Recognition System	• Identifies users for security or recommender	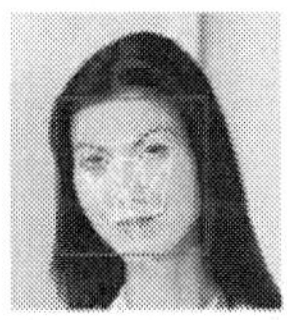
Smart Senior Corner	• Equipped with rehabilitation program • Interactive games with multi-touch display • Equipped with physical and smart living devices • Video chats with overseas friends & relatives	

Interactive Library Wall	• Interactive participation platform • Searching information in the Web • Used for meetings and interactive sharing	

4.1. The Library Is Intelligent and Organized

The attributes of "Intelligent" of the principle of Library3.0 are Self-renewing, flexible, functional, integrated, efficient, resilient, autonomous and sensitive (adaptive) via selective intelligence (wisdom of the expert), ontology, smart (intuitive) searching in natural language, and artificial intelligence while those of "Organized" are Turning the unorganized web of information into a systematic and usable body of knowledge by exhaustively describing and linking every piece of data to enable ease of access via semantic via Web/Search (Kwanya et al., 2013 [25] ; Noh, 2013 [26]). The two principles are interrelated with each other in reality.

Web3.0 enables automatic clustering based on similarity and relationships in the document content for easy searching. Some libraries are using Resource Description Framework (RDF), Ontology, Resource Description and Access (RDA) to organize and describe their collections. In the hardware side, the development of internet of technology (IoT) has impacted library models and its operation. The IoT library consists of information object, independent network and intelligent application. Information object refers to library literature resources. Its items include hardware facilities and security access control. The independent network denotes network systems with self-management abilities that have exhibit self-configuration, self-healing, self-optimization and self- protection capabilities. Libraries will offer learning experiences through self-serving and automation, efficient operation, user generated or personalized content, mobility and access 24/7. People of all ages will be able to experience and explore the world through convenient access to information and knowledge. Intelligent terminals help readers to use the library resources more efficiently, which therefore improves the resource utilization rates of libraries. It includes intelligent navigation retrieval terminal, security access control system, borrowing system and more (Wang and Zhao, 2015) [35] .

To start-off with RFID enabled facility including tag inside each book, self-check in/out kiosks, security gate, and RFID enabled book-drop for return, the library becomes self-service. More advanced kiosk is equipped with a detector that allows the station to adjust its height according to the user's height.

A 24-hour book drop along with a sorting unit improves the efficiency of the operation process. The device is able to efficiently and accurately sort the returned materials according to pre-defined categories. With the book drop and sorting unit, the workload of librarians is significantly reduced.

To make libraries even more efficient, smart bookshelves can be used. The smart bookshelf detects the on-shelf books automatically and updates the library management system database. The book list will then be generated on screen upon user enquiry at the input terminal at the side. The bookshelf also detects the book sequence, which allows for easy book searching. In addition, the mobile stock-take trolley can automatically detect the books on the trolley and informs the librarian of their positions on the shelf. The stock take trolley provides a book searching function and can identify what books are missing from the shelf after the stock-take with a handheld device. More advanced system such as Automated stored and retrieval system (ASRS) making use of robotic device is also available.

The booking system will then allow users to book library facilities such as function rooms and multimedia equipment. It will have customized ready scene settings and door access based on booking arrangements and more. Libraries can apply space on/of according to the user's access to enhance library automation functions. Within the function rooms, motion sensors will be installed to detect the flow of people in the area. Through identifying motion in the proximity, lights and air conditioners can be controlled automatically. When a user accesses the area, the lights and air conditioner will switch on. A light (lux) sensor will then adjust the brightness and temperature of the room accordingly depending on the traffic flow. The concept of "green energy" is not only practical but also educational to library users.

4.2. The Library Is a Federated Network of Information Pathways

The attributes of "Federated network of information pathways" of the principle of Library3.0 are supporting expressiveness and interoperability to create synergies between hitherto disparate information resources and systems via network of data (meaning), all-visible, all-accessible Web, and Cloud computing (Kwanya et al., 2013 [25] ; Noh, 2013 [26]). Information-discovery system has been adopted by some universities to provide access

7

CHAPTER

AN EMPIRICAL REVIEW OF LIBRARY DISCOVERY TOOLS

Xi Shi*, Sarah Levy

SUNY Rockland Community College, New York, USA

ABSTRACT

The Internet search concept has fostered an expectation that all users need to do is to feed relevant terms to a search engine to describe a topic or to ask a question, and click "search". The search engine is then expected to return a list of possible relevant and useful results for users to choose from. Based on this search concept, library system developers have been developing and constructing software programs for library databases that manage scholarly information. These software programs are known as library discovery tools, or web-scale discovery (WSD) tools. In this article, the term "library discovery tools" is used when discussing search engines designed for the libraries, WSD is included. Library discovery tools are intended for intelligent searches for educational or research purposes. This article provides a practical analysis of available library discovery tools in context of the present-day explosion of available open search engines on the Internet. The focuses of our analysis include how discovery tools are expected to manage library collections, provide access to scholarly information content, as well as other factors, such as budgetary considerations, when choosing or adding a discovery tool for a library.

Keywords: Information Search Engines, Library Discovery Tools, Information User Satisfaction, Information System Success

1. INTRODUCTION

Although the concept of library discovery tools is no longer novel to many, including librarians, the understanding of this tool is far from inclusive, nor is it conclusive as to what they do, how they do it, and the cost of delivering what is expected.

As both authors are librarians at a community college library, for the purposes of this article, we discuss library information management systems, library information search tools, particularly discovery tools, as IS products and services. Our library users, mostly our community college students, are observed and discussed as a specific sub-group of computing end users, or computing consumers. A library discovery tool is defined in this paper as a search engine that builds on unified indexes of licensed scholarly information, searches across multiple library databases provided by different vendors, and may be customized for its size, range and comprehensiveness of data inclusion for targeted solutions. Deliberations are presented on what search tools in academic libraries are most appreciated by college students and faculty, highlighting the differences between library search tools, including discovery tools, and commercial search engines.

Currently, the Internet is exploding with available free and fee based web-scale search tools for scholarly information, and many more are getting on Internet daily, of which some are commercial, such as Google Scholar, and others are sponsored by academies, such as Cite Seer X. Our analysis only includes those that are being commonly recognized and habitually utilized by college students for their comprehensiveness of academic information content coverage, and satisfying return of search results for discussion and examination.

A practical review is offered on how search tools are evaluated by users, and which tools are often identified and selected by specific clientele as appropriate for satisfying their information needs. First, a comparative analysis of commercial search engines versus library search tools is provided. Search experiences are discussed in order to describe characteristics of the various search engines, thereby identifying what information users expect from these search engines, and the factors users value when selecting one of many search engines for their information needs. Second, the most popular library databases and search tools are evaluated. The strengths and weaknesses are discussed to understand the practical functionality of each tool, and to facilitate the formation of realistic expectations of the currently available library discovery tools. Finally, financial cost-benefit considerations are reviewed as an additional factor to be weighed when library databases or search tools are being investigated.

2. COMMERCIAL SEARCH ENGINES

2.1. Google and Its Impact on the Library Market

Google, like it or not, has "created a model that librarians, as information providers, must meet head on." (Luther, 2003) [1] . It is in the best interest of librarians and library IS developers to understand Google, what it does, how it progressed to today's performance level and to learn from it, rather than distancing ourselves and our information users from it.

Most academic librarians do not, and generally speaking librarians should not, discourage students from using Google, especially Google Scholar, as Google Scholar never fails to provide relevant information for any set of search terms. Very often, Google Scholar includes full text information offered by educational institutions at no cost to the users. In some instances, the search will direct the users to their home library if the full text is found to be available via their library subscriptions. Even if the full text is not offered from the initial Google Scholar search, once the bibliographic information is provided, a targeted search by author(s) or title can be followed up in library databases rather easily.

Content retrieval from a Google search has no doubt been recognized by most educated users. Google's mega-data search algorithms, indexing preferences, ranking and displaying techniques have been observed, and its capacity of providing relevant contents, which is often described by many librarians as "good enough content" (Luther, 2003) [1] has been recognized and accepted. This is one reason that many library IS developers are imitating what Google does by creating "Google like" search tools to manage library information resources. However, one information retrieval characteristic presented by Google that has not received as much attention is that the relevant information content from a Google search is not always displayed directly or instantly. This statement may appear to be conflicting with many existing findings in the literature discussing how much information content Google does provide. We noticed Google often functions as a pointer, when it does not present or display full text information content. It does so sometimes even when it does deliver full text information content, so as purposefully to provide additional information from different sources. For example, when searching in Google for the term "define plagiarism", a seven digit result is indicated. The first entry listed on the result page is a definition of plagiarism provided by Google. Following this full text definition are links that point to definitions offered by other sources, e.g., dictionary.com, Wikipedia, and academic dictionaries, such as Oxford Dictionaries; as well as definitions provided by many educational institutions, e.g., Princeton University.

Many studies have reported and confirmed that academic library users use Google, love Google and complete their assignments on Google (Al-Maskari & Sanderson, M. 2011) [2] ; (Gross & Sheridan, 2011) [3] ; (Luther, 2003) [1] ; (Thompson, Obrig & Abate, 2013) [4]). We believe this user behavior is likely influenced by user's expectations. However, how the user's decision is made to turn to Google and return to Google still needs comprehensive investigations.

2.2. Search Engine Success Determinants

In the IS field, both researchers and practitioners, such as software developers, believe that user satisfaction is an indication of successful IS product design (Au, 2008) [5] ; (Bolton & Drew, 1991) [6] ; (Heinbokel, et al. 1996) [7] ; (Hsieh et al. 2012) [8] . This same belief is shared in library science (Condit Fagan et al. 2012) [9] ; (Gross & Sheridan, 2011) [3]). It is well researched that satisfied information users tend to continue with the system or programs they have experienced, and dissatisfied users will most likely look for different tools for the task at hand, or return to the tools with which they had prior positive experience. It is well documented that the influence of user satisfaction is closely associated with the selection of library IS products.

According to satisfaction researchers, consumer satisfaction is heavily influenced by consumer's expectations of a product or service being consumed. Therefore, many studies define satisfaction formation process as a comparison process in which consumers compare their pre-purchase expectations with the product or services after the purchase (Bhattacherjee, 2001) [10] ; (Churchill & Surprenant, 1982) [11] ; (Shi et al., 2004) [12] ; (Spreng et al., 1993) [13] , (Spreng et al., 1996) [14] . The most renowned theory that has been applied to explain information user satisfaction in both IS and library fields is disconfirmation of expectations. This theory proposes that consumer satisfaction is a result of a comparison of pre and post purchasing experiences. Before the purchase, consumers have expectations of what they are buying. During the purchasing process, consumers evaluate and select the products and services according to their pre-purchase expectations. After the purchase is made, consumers compare what they have purchased with their pre-purchase expectations. If the products or services exceed consumers' expectations, satisfaction occurs. If the products or services fall short of consumers' expectations, dissatisfaction occurs. In other words, disconfirmation is defined as the discrepancy between the consumers' expectations and the actual purchase experiences. Outcomes of the consumption that are better than expected lead to positive disconfirmation. Outcomes of the consumption that are worse than expected lead to negative disconfirmation.

As for information users, specifically library users who are consumers of IS products and services, satisfaction of search results is heavily influenced by their expectations. On one hand, it has been recognized that Google has changed information seekers' expectations, and defined and redefined the search experiences of those who are seeking information, scholarly and/or other kinds. On the other hand, library IS developers as well as librarians, have created very high users' expectations for library databases and discovery tools for scholarly information searching, promising discovery tools will do equally well, if not better than Google.

For the purpose of a comparison analysis, one search example is offered. Searching Google for the article Cognitive Test Anxiety and Academic Performance, by J.C. Cassady & R.E. Johnson in Contemporary Educational Psychology, the first listing shows the exact article with a brief description of this article. Clicking on the listing opens the article in Science Direct, and the full text can be retrieved in PDF. In addition, attached to the listing, users can easily find "Cited by" information, formatting guides to "Cite" in APA, MLA and Chicago. Or, user may choose "Save" the listing into "My library" for later review. Please note, Science Direct is one of our library's subscription databases, although no effort was put in from the library to have this article listed, and linked to the database by Google.

The same search is then conducted in Ebsco Discovery Service (EDS), which our library subscribes to. Searching by exact title, nine records are displayed. One of the nine records shows the exact title, authors and the publication information. However, the user is being pointed to Scopus, a database that our library does not subscribe to. If the search is revised to exact title and author(s), limiting to full text only, two records are listed. The record for the article from Scopus still shows, although no full text is available to our library users. The other record shows the title as "Regular Article: Cognitive Test Anxiety and Academic Performance", with a link indicating "View record from Science Direct", which will retrieve the full text article.

The above sample search demonstrates that the Google search probably exceeded the user's expectation, in a sense that the full text is quickly displayed in the search results, and can be retrieved with no additional effort from the user, and at no cost to the user. In addition, citation guides are handily provided for college students, and "Cited by" information offers immediate access to other research articles mostly in the same field where the original article was cited and discussed. On the other hand, the library discovery search tools, such as EDS, requires intensive customization from the library end, including constructing a "profile" to include all subscription databases, journals and other print and electronic materials. Regular updates of all holdings is required, including the loading of catalog MARC records if the library wishes their book collection to also be incorporated in the discovery search. Also, updating library collection information is the responsibility of the library, e.g., additions and deletions of journals and databases to the library collection. On the other hand, Google requires no effort from the user end to refer a search to a more appropriate database. All Google search features are integrated into one interface and any given search will be directed to the more appropriate search tools, which can also be activated at "More" and "Even more from Google". For example, if a search is detected as possibly being a scholarly article search, the search is automatically redirected to Google Scholar. Similarly, if a search is detected as likely being a book search, the search will go to Google Books. In addition, as Google announced in December 2009, it would begin to customize search results based on information gleaned from the user, indicating searches conducted by different individuals might receive different outcomes for their search results that are believed to better fit what those individuals have expected (Petter, DeLone & McLean, 2013) [15] .

3. USERS' PERCEPTIONS AND EVALUATIONS OF LIBRARY DISCOVERY TOOLS

In an information dynamic environment, consumers look for one stop shopping sites for their daily necessities, whether on the Internet or at local stores. Similarly, library users look for one-click search engines to fulfill their information needs.

The most widely recognized WSD services in the library market include Ebsco's Discovery Service, Ex Libris' Primo Central Index, Serial Solutions' Summon Service, and OCLC's WorldCat Local. Although these discovery tools are all web-based computer programs, the design of each was not based on the same conceptualization. The following discussion will explore some aspects that we believe are essential to understanding discovery tools in general, before attempting to select the most appropriate one for a specific library.

Ebsco's Discovery Services (EDS) is one of the discovery search tools highly recommended by librarians. Ebsco has won many academic contracts for the coming years. The State University of New York (SUNY), for example, signed a multi-year contract with EDS in 2013.

EDS obtains all licensed data from other suppliers. Wherever possible, EDS copies the bibliographic data and then includes it in Ebsco's metadata for processing and retrieval. By theory, this conceptualization should provide seamless searching across all information suppliers as well as direct retrieval without the use of pointers or linkers, since the data now resides with Ebsco. However, if any information or data is not included in Ebsco's central index, for example, if a supplier does not agree to supply their data, the discovery service will either omit the information from the search or it will need to rely on a linker to retrieve the full text. Another design conceptualization depends on common programming standards, which are also labeled in the field of library science as "discovery layers" (Hoeppner, 2012) [16] . Summon is an example of this design approach. Layers apply when indexing across different platforms used by different data suppliers for data retrieval. Although the interface, display, and searching process for this design approach may present as a single application to the users, it is in fact only an improved federated search adjusted by adding more layers to include more data access, under the condition that all suppliers comply with the same standards.

The following provides a description of the authors' evaluations based on our own trials and experiences at our library.

EDS, a rather new product by Ebsco in the library market, is developed and marketed as a "Google like" search engine. EDS is supposed to return information search results from all library resources across different platforms that are provided by different vendors. After the users enter keywords or subject terms for a topic, EDS would then perform a smart search to lead the users to results as well as offering suggested adjustments to the original search, such as narrowing or limiting the search. It may also provide recommendations for further exploration, e.g., providing different search terms, or searching in different areas. In addition, it can provide directions and assistance for further steps in the research endeavor. One example of such assistance is the citation guides that include most academically accepted citation formats. Ebsco has a solid reputation for their comprehensive collection content, sophisticated search options, and responsive technical support. However, with information content from other vendors included, such as Science Direct, LexisNexis, etc., the discovery module does not appear to be designed to return search results on an equal ranking across the providers. Some articles might be missing from the EDS search results, but would be retrieved if a search is conducted in the native database where the articles reside. This phenomenon confirms the bias of information retrieval as reported in other studies that confirm that the bias of search engine coverage and the bias of information retrieval system does exist (Buttcher & Soboroff, 2007 [17] ; Vaughan & Thewall, 2004 [18]). Consequently, users must understand that the search results provided by a discovery tool from one specific vendor may not be content neutral, whether it is due to the indexing technique or the classification standards. In other words, the Ebsco Discovery Service user should expect to be pointed to Ebsco information first and foremost.

From our experience, help from librarians often enhances the EDS search outcomes. As a matter of fact the librarians' assistance, which might include subject phrasing choices, searching technique guidance, is found most effective as library users appreciate librarians comprehensive knowledge not only regarding the subject matter of the information content, but also the performance of library subscription databases. Unlike Google searching, librarians are expected to be familiar with the content specialty of each database, search functionality of different vendors, as well as being knowledgeable about their institution's academic curriculum. EDS, although a discovery tool, works better with librarians' intervention matching the available information to satisfy the users' needs, in our case mostly students' information needs to complete their academic assignments. For example, librarians would recommend information from Science Direct as an intelligent choice for scholarly information on "joint injury", but for a specific legal case involving "joint injury" the user would receive better results in LexisNexis.

To better appreciate the consequences of the design concept in understanding how library discovery tools might function differently from one another, and how they differ collectively from commercial search engines such as Google, the following offers a factual analysis from our observations as librarians. The discussion of application evaluations and the users' selection of search tools are presented from three perspectives. First, the content orientation of the provider; second, users' information needs, information literacy level and their perceived net benefits, and third, budgetary or resource-based considerations and their influence on selection of a search engine, or search tool.

3.1. Content Orientation of Search Engine Providers

First, we should acknowledge that every library database has its own content orientation, in much the same way that commercial search engines are constructed. Although Google can be viewed as a discovery tool, when specific information is being searched for, the Google user will be redirected to a more appropriate search engine. Sophisticated users may directly choose a most effective engine for their particular search. For example, when looking for the definition of a term, a user may go directly to a dictionary or encyclopedia, or they might try Google first, and then be redirected to a more appropriate site. Most current generation college students know to go directly to Amazon when seeking to purchase books, to Wikipedia for quick information on a subject, and to their college website for academic related information, such as registration dates, an academic calendar or course information.

Library databases or search tools also have their own specialties in content collections. Therefore, the database construction concept of each provider may be influenced and governed by the distinctive knowledge of their collections, and created to work most efficiently with their own collections. For example, Gale specializes in literature; Ebsco started their library business with journal and serial publications; and OCLC is known as a bibliographic utility. Understanding the specialties of each vendor is important because it lands users in a position to form realistic expectations. For example, if users recognize that Gale specializes in literature content, they may reconsider if the use of a discovery tool is most sensible to find literary criticism, even if the Gale content is included in the discovery module. Understanding that

began as a journal servicer, users can expect their serial data collection to be one of the most inclusive. Similarly, OCLC's WorldCat should provide superior functionality when searching catalogs for library specific holding information, resource sharing possibilities, etc. Periodical or journal information at the publication level provided by OCLC derives from MARC records entered or loaded from other sources. Consequently the information quality may not be predictable, because the host of such information content may or may not behave as described. Similarly, Ebsco may not be the best tool for searching library catalog information, especially union catalogs, since this information is loaded in MARC format from other sources, such as OCLC or a local library system, into the EDS module. Anytime data is introduced from an external source, uncertainty can be anticipated, and users may experience less reliable search outcomes, which will inevitably lead to less confidence in the searching process.

The following scenario offers two search examples to illustrate the impact of the selection of a search engine on a search outcome. Students in a social studies class are assigned to research the correlation between quality and the cost of higher education. The results from different search engines are listed in Table 1 below for analysis. The same search term "higher education cost and quality" was applied to all searches.

The data (retrieved July 2015) from Table 1 shows the search tools selected, the search strategy applied and the results retrieved from the initial search conducted when the assignment was received. Please note that of the 65,746 EDS results, 36,827 are full text articles and all Science Direct listings have full text. The Google Scholar search was not initiated by the user. Instead, the Google Scholar search results were pointed to by Google web.

We conducted this same search in EbscoHost with Academic Search Complete, Education Source and ERIC simultaneously selected. The search returns 4813 results and 1488 are displayed when limiting these results to full text scholarly journals. See Table 2 for details of this search.

After the authors' careful evaluation of the results from this search, we found that the EbscoHost search results displayed in Table 2 provided much more focused content than the same search performed in the Ebsco EDS, which, like most library databases, has the ability to suggest subject terms within the search context to narrow down the results. Most databases also build in the "smart search" functionality to suggest similar or broader terms for expanding a search.

Our students, as is the case for most undergraduates, prefer to spend the least possible time on receiving or retrieving a minimum amount of acceptable information (Gross & Sheridan, 2011) [3] ; (Hoy, 2012) [19] . This behavior will be further discussed in later sections of this paper. From the analysis of our sample searches, two indications as to why the selection of a more effective search tool is important warrant attention. One indication is that a wise selection of a search tool can result in speedy and satisfactory information retrieval, creating an immediate positive perception by users of the search tool for their information search process. Second, this positive perception will most likely be reinforced by repeating positive experiences when selecting an effective search tool. In turn, the positive experiences enhance users' learning practice of evaluating available search tools, and then selecting the best one that may satisfy their information needs.

The following sample searches are provided to further analyze the importance of understanding the impact of selection of an effective search tool. Searching in Google for "Wharton Puts First-Year MBA Courses Online for Free" published by Business Week, September 13, 2013, pulls up the full text article. The same search was conducted in Ebsco EDS, EbscoHost, ProQuest Central, and Gale. No result was found from any of the tested databases, nor was the user pointed to other site(s), although Business Week (also indexed as Bloomburg Business week and Businessweek.com) is covered by Ebsco Business Complete to which our library subscribes, and it is also included in our individual journal subscriptions. The retrieval failures of newspaper articles, including from major papers such as The New York Times and The Wall Street Journal, are often found from the library databases, including discovery tools. Excuses from the providers include the incompatibility of the content indexing from the dot com version and print paper, misinformation received from the newspaper(s), etc.

3.2. Information Users Needs and Perceived Net Benefits

To begin our discussion on users' information needs and the role those needs may play in information searching and retrieval, a brief literature review is provided.

Table 1. Search result comparison.

Search Engine	Search Method	Search Results
Ebsco EDS	Keyword(s)	65,746
Science Direct	Default	202,970
Google	Default	140,000,000
Google Scholar	Default	4,400,000

Table 2. Ebsco education research complete, academic search complete & ERIC.

Search Engine	Databases	Search Method	Search Results
		Keyword(s)	4813
EbscoHost	Education Research Complete Academic Search Complete Eric	Limit to full text only	3262
		Also limit to scholarly journals	1488

Many satisfaction researchers in the field of IS and marketing consider it necessary to include a "needs" construct when studying information searching and retrieval. Findings of the theory development and test results were reported (Oliver, 1995) [20] ; (Spreng et al., 1996) [21] . Needs theory can be traced way back to the 1940s, when Maslow (1943) developed the theory of a hierarchy of needs. The basic assumption of needs theories indicate that when deficiencies of need are realized and identified, people are motivated to take actions to correct or reduce those deficiencies to satisfy the need. Since human beings have different levels of needs according to Maslow, actions taken would indicate different behavior patterns in an effort to fulfill those needs. Two areas of concern are worth noting. First, many satisfaction researchers argue that needs should be one of the driving factors in understanding consumer behavior, and that needs is likely to have some effect on predicting or influencing consumer satisfaction. Therefore, the construct of needs should be included into the disconfirmation equation. In doing so, some researchers find that consumers may compare the product performance, such as IS programs or search engines, with their identified needs. Accordingly, if the product performance exceeds the innate needs, the consumer realizes the personal benefits, thus is satisfied. On the other hand, if the product performance does not fulfill the pre-identified needs, for example, the search failed to provide the sought after information, the consumer would perceive no net benefits, thus is dissatisfied (Petter et al., 2013) [15] ; (Sirgy, 1984) [22] ; (Spreng et al., 1996) [14] ; (Shi et al., 2004) [12] .

Researchers in the field of information science noted another perspective on the impact of needs on satisfaction. This perspective denotes that unlike other human needs, such as the need for food, water or shelter, etc., "what is required to satisfy an information need is often not known to the individual concerned" (Cole 2011, p. 1216) [23] . Information needs must be placed in context of the specific user's situation in order to be meaningful. Saracevic (2007) [24] stated that while most research on information needs concerns the user-centered concept, computer scientists when designing information retrieval systems often ignore the information-situation condition. Because the nature of information needs is intangible and non-specifiable, the information users may be engaged in a long process of activities as they search and then revise their searches to find the needed information. Therefore, no matter how long the process of information searching takes and how it progresses, e.g., an academic assignment or school project may in fact last weeks or months, there will be different types of information searches throughout that time while the information needs for this academic assignment remain constant. This information seeking and searching behavior can be witnessed in libraries with students looking for information to satisfy their assignment or research project. Very often students come in to the library without knowing or understanding what information they may need, or where to start searching for the information that may fulfill their assignments. Over the course of doing and completing their assignment, students learn to recognize and better understand their information needs, and the information needs become clearer and clearer as the information seeking activities continue. It is a learning process, as students conduct information searches trying and using different tools to get and review various search results. This learning process helps the students realize and clarify their information needs. In the meantime, their experience with searching activities and the applications of those tools in their searches, forms expectations of what they can find, which tools are effective, where they should start, and how to get the most needed information in the shortest time.

Additional theories on how users' needs should be integrated into IS design warrants further review. Users' information needs have been widely studied and recognized by researchers in many different fields. However, how the needs are understood and integrated into the IS design is approached differently. The concept of cognitive approach is attention worthy. The cognitive approach considers that it is the information specialists' knowledge that determines the successful design of an effective information retrieval system, and that knowledge is not based on the empirical studies of users (Hjorland, 2013) [25] . Examples of this approach include Apple computers and iPhones. Apple's success suggests that understanding what users may want or need as individuals, as well as socially and culturally, offers insight for their design. Apple's success demonstrates that even if a company does not interact directly with users, even if the design is not based on a review of the market, it can still be much more insightful about what people could want (Verganti, 2009) [26] . The library catalog as an IS retrieval system is a similar example. The most

IMPLEMENTATION STAGE: INTEGRATING WITH ELECTRONIC MAPS

Contents

Chiayi, a historical city in Taiwan, has played an important role in the past hundred years of Taiwan history. Thematic information about Chiayi includes language, history, geography, nature and arts. Data related to Chiayi City is identified in some of the archives, which consists of different types of data including images, texts, maps, drawings and sound collections. To reduce the waiting time for visualization, a client-server architecture was implemented. The front-end interface used Flash technology and the web server stored the archive data.

Design

GIS provides the users with intuitive perception through information visualization. Atlases and photos describe the landscape created by either natural processes or human activities during a period of time, annotate what happened during specific space-time conditions, and reflect the culture of a place (Summerby-Murray, 2001). Reviewing maps of the same place from different periods can assist users to rehabilitate history and understand changes in the environment, society and culture. In order to correlate unstructured information such as images, sounds, drawings and textual descriptions with spatial information, the system designed three levels of interaction to represent the complex associations: an interactive map, a time map and a hyperlinked map. There were two main user interface designs to facilitate learning.

The first component was an interactive map. Layers of maps, which are similar to GIS, were used as the information visualization technology to organize and display various kinds of information for every point on a computerized map. This involved powerful, complex computer databases that organized information around a specific location (see Fig. 1). Users could select a thematic representation or an automatic combination of layers of maps. The thematic representation was selected and tested as a better view to comprehend the information about a specific topic about Chiayi (see Fig. 2).

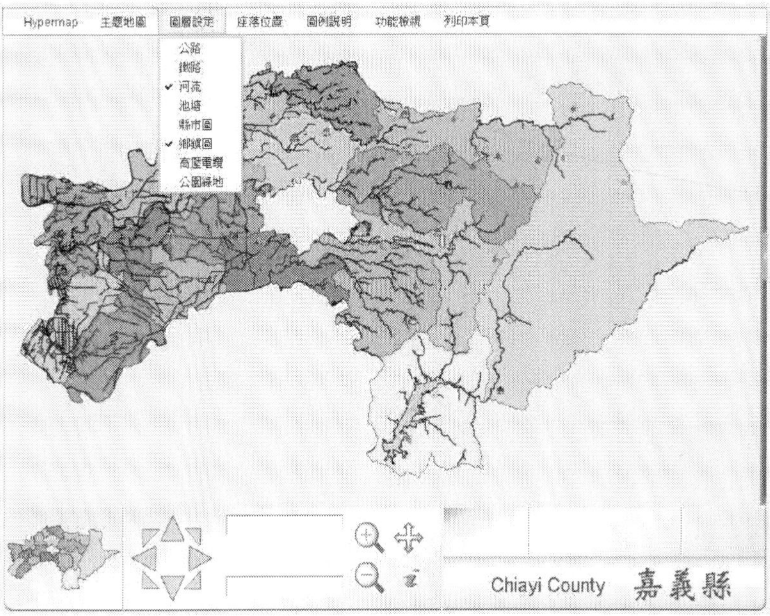

Figure 1. Users can zoom, move and select different map layers

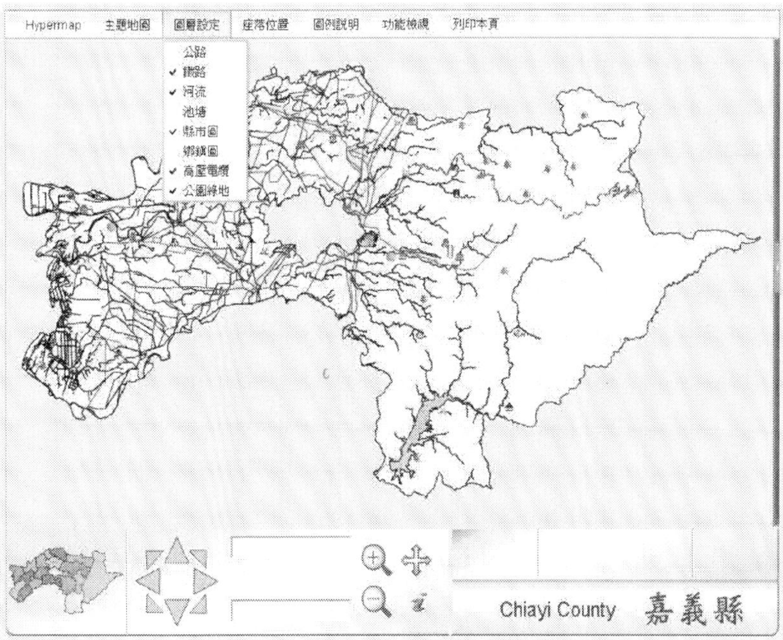

Figure 2. Users can select map layers to obtain a better view about a specific topic

The other design was a time map. Three major challenges for a time-series data exploration system are providing algorithms for the analysis and creation of metadata, filtering out data that is uninteresting, and the interactive exploration of the regions of interest. In this project, a time map was used a tool to filter out uninteresting data (Grady, Flanery, Donato, & Schryver, 2002). Sliders, which are a generic user input mechanism for specifying a numeric value from a range, were used to control a threshold filtering the entities shown on the display. In this level of interaction, the time map provided users with a tool to move between different representations, thereby enabling them to explore the data from several perspectives. The data visualization slider, designed as a chronicle scale, was tied to a set of map layers. A user operated the slider by holding down the left mouse button and moving to a new position. Information within the time interval would be displayed on the map (Figure 3). In addition, a hyperlinked map was designed to retrieve further information. A straightforward approach of enabling users to dynamically retrieve time–sensitive information is to link further information to the entities distributed over the map, based on the values in the chronicle field (Risch, et al., 1997). Users were able to click on each entity shown on the map and retrieve additional information through a pop-up window that appeared. By clicking on keywords on the windows, the hyperlinks would display more information stored in the original archives (Figure 4).

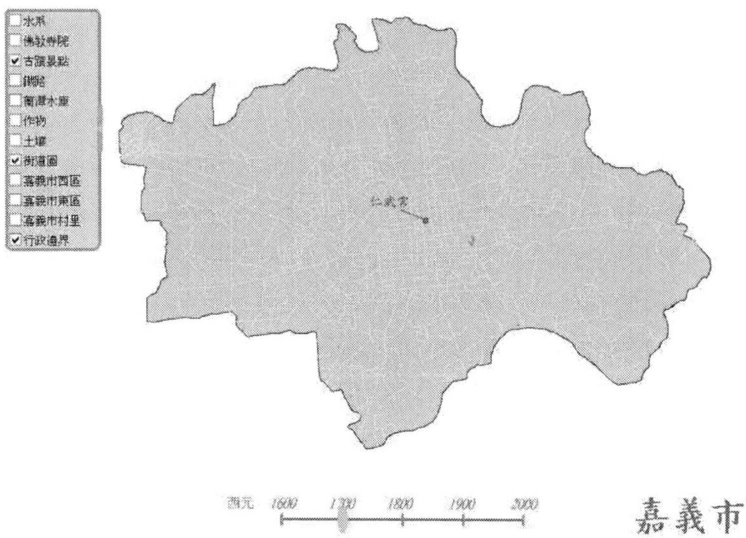

Figure 3. The slider sets the calendar year and related information is shown.

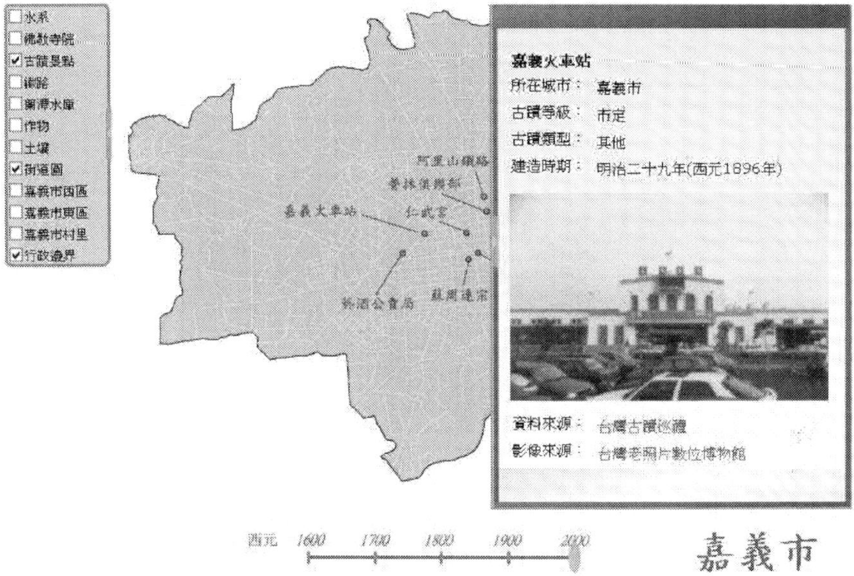

Figure 4. A pop-up window and hyperlinks retrieve information.

Instruction and evaluation

To evaluate the effects of these learning environments, two empirical studies were conducted in two elementary schools respectively (see Figure 5). In the first study, the subjects were 32 fourth-grade students. The teacher integrated the electronic map and digital archives with classroom instruction. The achievement test after instruction showed that the students performed significantly better than those under traditional instruction. In the second study, project-based learning activities were arranged to collect qualitative data for assessment. After becoming familiar with the resources, the students were divided into several groups to complete their own project. The project for each group was to construct a thematic map based on the electronic map and digital resources. Students in each group participated in the activities of discussing the theme, distributing tasks, collecting data, constructing the map and preparing a slide to present their works. Analysis showed that the students possessed a positive attitude toward using digital libraries for constructing their own presentation. In addition, the teacher was convinced of the educational advantages of using digital archives. Through watching the students working together as a team, she realized that the digital archives were playing the role of mediating the students' active learning. In addition to factual knowledge, the students also learned by doing.

Figure 5. Two teachers using digital libraries in classroom instruction

PROMOTION STAGE: ADDITIONAL MEDIA AND WORKSHOPS

Extension and promotion

The successful experience encouraged the research team to promote additional learning resources. Resources on Chiayi City and Chiayi County were developed in the previous stage. However, there are 16 cities or counties in Taiwan. The research team utilized the same design pattern to create resources on the other 14 cities or counties (see Figure 6). Later, several training workshops were held to train teachers how to use the digital libraries. These teachers were first introduced to the concept of digital libraries and their educational applications. Then, the speaker demonstrated how to use a web-based interactive map to retrieve resources related to the curriculum. At the end of the workshops, questionnaires were dispatched to assess their impressions and intention to use these materials. Table 1 shows the workshop attendants' positive attitude toward these materials.

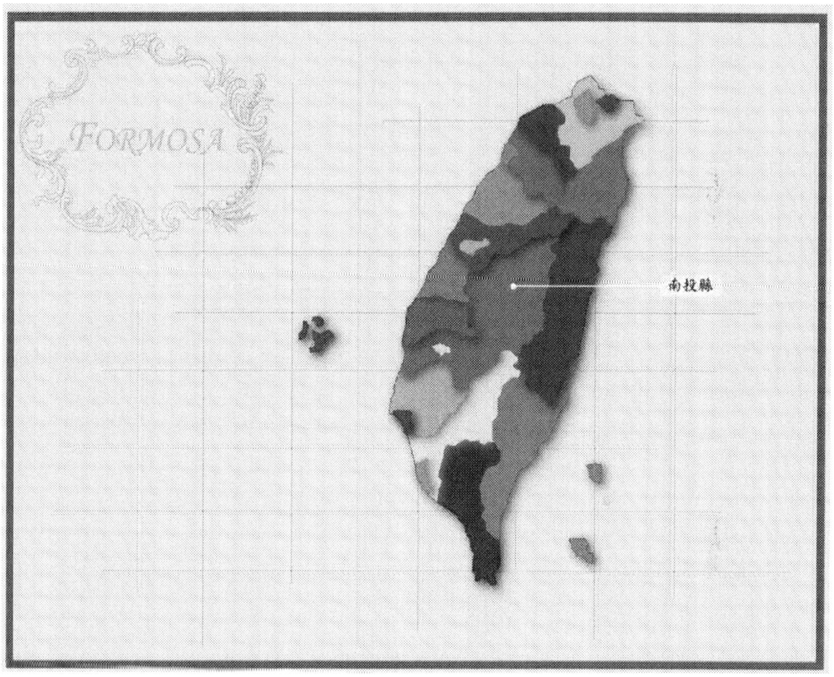

Figure 6. Users can choose specific cities in Taiwan when accessing the digital library.

Table 1. Workshop attendants' attitude toward the developed materials (N=41)

Items	Strongly agree	agree	neutral	disagree	Strongly disagree
This web-site is well designed	12	26	3	0	0
This website will help to comprehend the contents	22	15	4	0	0
You will use this website in your instruction	13	23	4	1	0

Follow-up evaluation

Six months after the workshops, the research team interviewed the attendants on the phone to understand the status of their digital library utilization. Surprisingly, none of the interviewees actually used these resources in the classroom. The major reason for not using these media was a lack of time. Since the textbooks used in the classroom were not city- oriented, teachers had to spend time retrieving related information for certain cities. Teachers were too busy to re-organize these materials for instruction. If the resources could be designed as independent components according to the concepts in the textbooks, teachers could easily select related media and assemble them as teaching resources.

ENHANCEMENT STAGE: REVISION FOR TEACHERS' NEEDS

In light of the importance of providing easy access to media for instruction, the research team selected themes on domestic geography in the social studies curriculum as the teaching content and designed computer animations as the instructional media to foster teachers' interests in integrating technology with classroom instruction. There were two major enhancements: one was to organize the resources into the structure of the textbooks, and the other was to organize the resources into themes.

Resources in textbook structure

The goal at this stage was to bridge the gap of expected usability between the instruction material developers and the practicing teachers. From the perspective of the teachers, integrating the GIS and digital archives with the existing curriculum as packages would help teachers to use these instruction materials. The tasks of this project included selecting related units in the textbooks, searching for useful resources in the digital archives, and then designing instruction plans and digital resources. Table 2 shows the selected units and the associated media.

Table 2. Textbook units selected for media development

School year	Unit Name	Number of Lessons	Media
Grade Four	Name and location of hometown	3	Interactive maps
	Natural environment and living of hometown	2	animations
	Development of hometown	2	animations
	Festivals and folk cultural activities in hometown	2	Video clips
	Unique sights and products in hometown	2	pictures
	Tours around hometown	2	maps
Grade Five	Where is Taiwan	2	maps
	Natural environment of Taiwan	3	Satellite images
	Resources in Taiwan	2	pictures
	Population and change of towns	2	Interactive maps
	Area and traffic	3	Interactive maps
	Care for Taiwan	2	pictures
Grade Seven	Environment of Taiwan	6	pictures
	History of Taiwan	6	animations

Resources in theme structure

The themes related to geographic subject content in the elementary and junior high school social studies curriculum were associated with the topics of population, economy, settlement and traffic, as well as regional development and environmental protection. With the combination of texts, pictures, digital maps, simulated animations and games, these e-learning resources supported teachers in constructing a learning environment to engage students in learning and discussion. Animated simulations were used to illustrate abstract concepts. For example, an animation with different phases of urbanization aroused students' feelings towards various environments, and provided students with a context to share their experiences (see Fig. 7). This was helpful for teachers to explain the process of how a settlement develops in the suburbs. As Brookfield (1987) pointed out, the process of internally examining and exploring an issue of concern, when triggered by a relevant experience, creates and clarifies meaning in terms of self and results in a changed conceptual perspective.

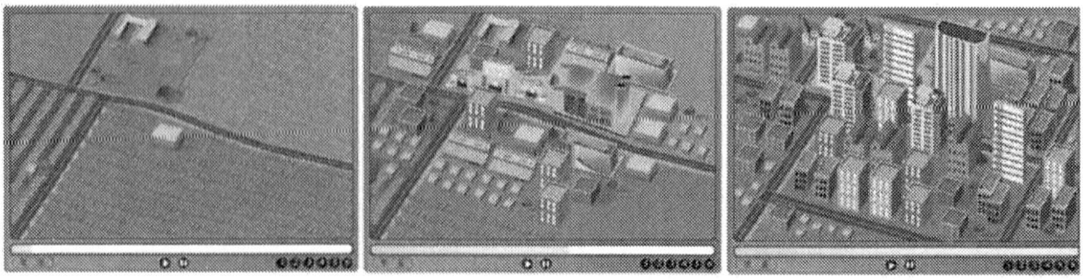

Figure 7. Simulation of the urbanization process. From left to right: Phase 1, 3 and 6.

Thematic maps were used for cross-referencing to create meaningful learning. Animations provided cross references associated with a particular topic, to encourage students to reason and seek evidence for arguments. To develop their critical perspective, students need to reason within various points of view and use evidence in order to draw conclusions, make decisions or seek solutions. For example, to answer why the intensive construction of traffic infrastructure was needed on the west coast, a link to a population distribution map and a traffic network map provided evidence for reasoning (see Fig. 8). In addition, statistical data could be converted into a graphical form to facilitate comprehension (see Fig. 9).

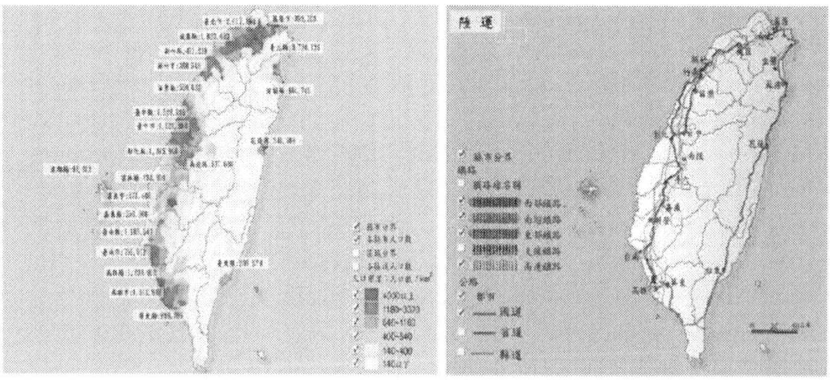

Figure 8. Maps referencing traffic issues. The left is the population distribution and the right is the traffic network.

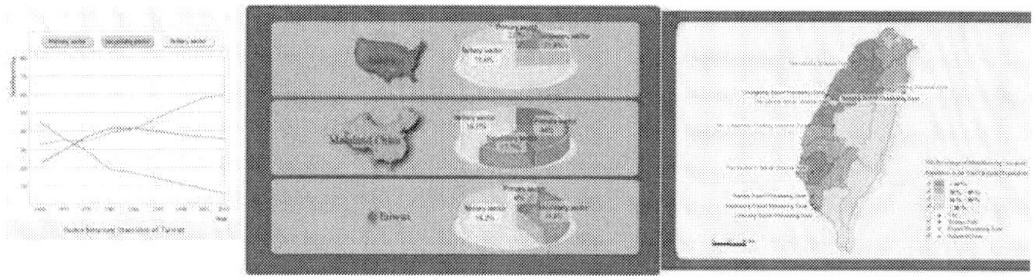

Figure 9. Visualization of economic transition and distribution of industrial parks

Figure 10. Settlement designs by two students.

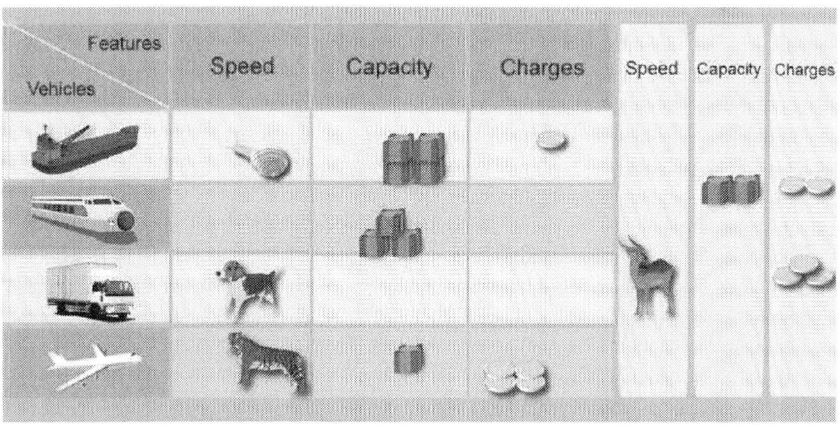

Figure 11. Interactive matching quiz partly completed by a student

FUTURE WORKS

The future work is to develop digital media and activities on interactive whiteboards based on the digital archive resources. The research team will first search and organize the resources in the digital archive project, and then integrate in a spatial concept with touch- control operations to design efficient instructional activities and media. For example, as shown in Fig. 10, an interactive tool for settlement planning was developed for students to practice their urban design. Using drag and drop operations, students arranged constructions such as buildings or airports according to their ideal locations to build a village or city. Students' works were then be presented and evaluated by classmates. Students with diversified backgrounds and values designed cities with a variety of features and functionalities. The teacher used challenging questions to inspire students to respond, expand and develop the topics during critical–talk lessons. In addition, an interactive quiz to compare traffic infrastructure was used to attract students' attention and efforts on reaching consensus (see Fig. 11). According to Moon (2008), peer assessment provides practice in making judgments on the basis of evidence. The act of assessing the work of another is a matter of making a judgment. It more deeply involves learners in the process of meta- cognitive critical thinking skills.

CONCLUSIONS

Web-based instruction is known for its media-rich online environment, providing users access to remote resources for self-paced learning. Although a number of schools have designed educational websites to integrate in-classroom learning activities with the school curriculum, designing effective learning resources based on sound educational theory will encourage students and teachers to use them more often.

Constructivist educational models are based on the theory that instead of being passive receivers of information, learners should be active explorers of their own understandings. Teachers and technology merely serve as mediators or guides to support the development of learners. Several researchers have treated technology as a vehicle to foster active learning and believe that such learning environments motivate and facilitate the acquisition of knowledge by providing an intuitively comprehensible context (Milson & Earle, 2007; Papert, 1980; Piburn, Reynolds, MacAuliffe, Leedy, & Birk, 2005).

The prevalence of computer usage in schools is driving the need to understand its effects on learning when technology is integrated within instruction. The experiences of this project addressed the needs of teachers to enhance students' geographical knowledge through the use of GIS-based resources. Such knowledge is a useful reference when examining issues related to educational practices using digital libraries.

REFERENCES

1. Ausubel, D. P. (1978). *Educational Psychology: A Cognitive View*. New York: Holt McDougal.
2. Brookfield, S. (1987). *Developing critical thinkers: Challenging adults to explore alternative ways of thinking and acting.* San Francisco: Jossey-Bass.
3. Bunch, R. L., & Lloyd, R. E. (2006). The cognitive load of geographic information. *The Professional Geographer, 58*(2), 209-220.
4. Eveland, W., & Dunwoody, S. (2000). Examining information processing on the WWW using think aloud protocols. *Media Psychology, 2*(3), 219-244.
5. Frias-Martinez, E., Magoulas, G., Chen, S., & Macredie, R. (2006). Automated user modeling for personalized digital libraries. *International Journal of Information Management, 26*, 234-248.
6. Grady, N., Flanery, R., Donato, J., & Schryver, J. (2002). Issues in time-series and categorical data exploration. In U. Fayyad, G. G. Grinstein & A. Wierse (Eds.), *Information Visualization in Data Mining and Knowledge Discovery* (pp. 229-235). San Francisco: Morgan Kaufmann.
7. Milson, A. J., & Earle, B. D. (2007). Internet-based GIS in an inductive learning environment: A case study of ninth-grade geography students. *Journal of Geography, 106*(6), 227- 237.
8. Moon, J. (2008). *Critical thinking*. New York: Routledge.
9. National Research Council (2006). *Learning to think spatially: GIS as a support system in the K- 12 curriculum*. Washington D.C.: National Academies Press.
10. NDAP (2003). NDAP Introduction,

Accessed at http://www.ndap.org.tw/1_org_en/introduction.php

11. Nilsson, R. M., & Mayer, R. E. (2002). The effects of graphics organizers giving cues to the structure of a hypertext document on users' navigation strategies and performance. *International Journal of Human-Computer Studies, 57*, 1-26.
12. Papert, S. (1980). *Mindstorms*. London: Harvester Wheatsheaf.
13. Piburn, M. D., Reynolds, S. J., MacAuliffe, C., Leedy, D. E., & Birk, J., P. (2005). The role of visualization in learning from computer-based images. *International Journal of Science Education, 27*(5), 513-527.
14. Risch, S., Rex, D. B., Dowson, S. T., Walters, T. B., May, R. A., & Moon, B. D. (1997). The STARLIGHT information visualization on system. *Proceedings of the IEEE Conference on Information Visualization*, 42-49.
15. Schwartz, N. H., Verdi, M. P., Morris, T. D., Lee, T. R., & Larson, N. K. (2007). Navigating web-based environments: Differentiating internal spatial representations from external spatial displays. *Contemporary Educational Psychology, 32*(4), 551-568.
16. Slocum, T. (1999). *Thematic Cartography and Visualization*. Upper Saddle River, NJ: Prentice Hall.
17. Sui, D. Z. (2004). GIS, Cartography, and the ? hird Culture?? Geographic Imaginations in the Computer Age. *Professional Geographer, 56*(1), 62-72.
18. Summerby-Murray, R. (2001). Analysing heritage landscapes with historical GIS: Contributions from problem-based inquiry and constructivist pedagogy. *Journal of Geography in Higher Education, 25*(1), 37-52.

9
CHAPTER
SECURITY AND DIGITAL LIBRARIES

Edward Fox and Noha ElSherbiny

Virginia Tech, USA

INTRODUCTION

Security is an important issue in digital library design. Security weaknesses in digital libraries, coupled with attacks or other types of failures, can lead to confidential information being inappropriately accessed, or loss of integrity of the data stored. These in turn can have a damaging effect on the trust of publishers or other content providers, can cause embarrassment or even economic loss to digital library owners, and can even lead to pain and suffering or other serious problems if urgently needed information is unavailable (Tyrväinen, 2005).

There are many security requirements to consider because of the variety of different actors working with a digital library. Each of these actors has different security needs (Chowdhury

& Chowdhury, 2003). Thus, a digital library content provider might be concerned with protecting intellectual property rights and the terms of use of content, while a digital library user might be concerned with reliable access to content stored in the digital library. Requirements based on these needs sometimes are in conflict, which can make the security architecture of a digital library even more complex.

The design of the security architecture of a digital library must go beyond simply adding one or a few modules to a previously designed system. This is because there may be security holes in pre-existing modules, and because difficulties can arise when attempting to integrate the modules. The security architecture of a digital library must be designed so that security concerns are handled holistically. A security system designer must view the whole architecture and consider all of the applicable security factors when designing a secure digital library. The nature of a security attack may differ according to the architecture of the digital library; a distributed digital library has more security weaknesses than a centralized digital library.

Security attacks can be categorized as physical attacks and logical attacks (Stallings, 2006). A physical attack involves hardware security where keys, locks, cards, and visitor monitoring is used. A logical attack involves an attack on the content or digital library system. We focus on the logical attacks and software security of digital libraries.

SECURITY ISSUES WITH DIGITAL LIBRARIES

According to the DELOS Reference Model (Candela et al., 2007) there are 6 main concepts in a digital library universe: content, user, functionality, architecture, quality, and policy. Each of these concepts has security issues that affect it.

2.1 Content

The content of a digital library includes the information objects that a digital library provides to the users. Some of the security issues involved are integrity and access control. Integrity requires that each object/resource has not been altered or changed by an unauthorized person. Access control encompasses two security requirements. The first is authentication where the user must log into the system while the second is confidentiality, which means that the content of an object is inaccessible by a person unless they have authorization. Not all digital libraries are free; often content is provided to digital library users for a certain fee, whereupon access control is needed to protect the content. Further, some content is inappropriate for some users, or targeted to particular user groups; there are a whole host of such other reasons for access control.

Logical attacks such as hacking and message tampering can affect the integrity and confidentiality of the content. Improved information access in digital libraries has raised many issues that affect the management of digital libraries. Content Management, or more specifically Digital Rights Management, refers to the protection of content from the different logical security attacks and issues relating to intellectual property rights and authenticity.

2.1.1 Digital rights management

DRM provides content protection by encrypting the content and associating it with a digital license (Tyrväinen, 2005). The license identifies the user allowed to view the content, lists the content of the product, and states the rights the user has to the resource in a computer readable format using a digital rights expression language (DREL) or extensible Rights Markup Language (XrML) that also describes constraints and conditions.

There are 7 technologies used to provide DRM (Fetscherin & Schmid, 2003). Table 1 summarizes the DRM components and supporting technology.

Each of these components involves mechanisms used to provide DRM:

- **Encryption**: Encryption techniques such as symmetric and asymmetric ciphers can be used to provide access control; public-key encryption is used in payment systems that control how and by whom the content is used.
- Symmetric ciphers using DES, 3DES, AES, and RC4 algorithms require the use of a shared secret key to encrypt data before it is sent. At the receiver's end the cipher text is decrypted using the same secret key. Symmetric ciphers depend on both the sender and receiver knowing the shared key.
- Asymmetric ciphers use a pair of keys, public and private, for each of the sender and the receiver. The public keys of both the sender and the receiver are known but the private key is kept secret. If encryption is performed using the public key then only the private key can be used for decryption and vice versa.
- **Passwords**: Stored strings must be matched by users desiring access.
- **Watermarking**: Characters or images are added to reflect

ownership. Steganography is used to conceal data inside audio, video, or images (Johnson & Jajodia, 1998). Different watermarking techniques have different aims; some watermarks might be visible while others invisible. Some watermarks are reversible (Mintzer et al., 1997); it depends on the desired use of the watermark and what is being protected.

- **Digital signature**: Asymmetric encryption can be used. Likewise, hash algorithms such as MD5 and SHA can be used to create a signature (Stallings, 2006).

Table 1. DRM Components and Protection Technologies, adapted (Fetscherin & Schmid, 2003)

Component	Protection Technology
Access and usage control	Encryption (e.g., symmetric, asymmetric), passwords
Protection of authenticity and integrity	Watermarks, digital signatures, digital fingerprints
Identification by metadata	Allows description of an object in suitable categories, covering the digital content, rights owner, and conditions.
Specific hardware and software	Includes all hardware and software used by the end-device through which the digital content is being played, viewed, or printed.
Copy detection systems	Search engines, which search the network for illegal copies and use watermarking.

Payment systems	Can be seen as a certain type of protection technology as it requires user registration, or credit card authentication, which also require a trust relationship between the content provider and the customer.
Integrated e-commerce systems	DRMS must include systems which support contract negotiation, accounting information, and usage rules.

- **Digital fingerprint**: Digital fingerprints are a more powerful technique involving digital signatures and watermarking. The creator of the content creates a unique copy of the content marked for each user; the marks are user-specific hence called fingerprints. Should a user illegally distribute the content, the creator can use search robots to find those copies (Schonberg & Kirovski, 2004).
- **Copy detection systems:** Search engines also can help locate such copied objects. Copy- detecting browsers can protect digital content too.
- **Payment systems:** Users must divulge personal information to pay for content. Installing payment systems can help protect digital content.

There is no standard mechanism for providing DRM, mainly due to the lack of regulations (Chowdhury & Chowdhury, 2003), however there are various systems and protocols introduced to provide content management and support fair usage policies.

There is a tradeoff between security and performance. Nadeem and Javed use a Pentium-4,

2.4 GHz machine running Microsoft Windows XP operating system, encrypt 20527 bytes to 2323398 bytes of data using DES, 3DES, and AES. For 20527 bytes of data it took 2 seconds to encrypt using the DES algorithm and 4 seconds to encrypt using the AES algorithm (Nadeem & Javed, 2005). It can be seen that the more complex the encryption algorithm the longer it takes to encrypt the data. In another study, encrypting data with the RSA algorithm using a key size of 1024 took 0.08 milliseconds/operation on an Intel Core 2 1.83 GHz processor under Windows Vista in 32-bit mode, while using a key size of 2048 took 0.16 milliseconds/operation (Dai, 2009).

User

The User in a digital library refers to "the various actors (whether human or machine) entitled to interact with digital libraries" (Candela et al., 2007). Digital libraries connect the different actors with the information they have and allow the users to consume old or generate new information. Security issues relating to the users of a digital library intersect with content issues discussed above. A main logical security issue relating to users and content is access control. Different access control requirements arise for distributed systems (Tolone et al., 2005) to ensure both confidentiality and authentication:

- Access control must be applied and enforced at a distributed platform level, so should be scalable and available at various levels of granularity.
- Access control models should allow a varied definition of access rights depending on different information and must be dynamic where changes to policies are easily made and easy to manage.
- "Access control models must allow high-level specification of access rights."(Tolone et al., 2005)

Digital library users may need to be authenticated before they can access content. Global/universal identification may not suffice. A service provider that provides content based on a non-identity based criteria like age will not benefit from global identification because there is no way to verify the authenticated user's personal information. Usernames and passwords are not efficient ways to provide authentication.

One of the most widely used authentication protocols is Kerberos. It (Neuman & Ts'o, 1994) is a client–server model, which secures communication with servers on a local network. Developed at MIT in the 1980s to provide security across a large campus network, it is based on the Needham-Schroeder protocol and has now been standardized and included in many operating systems such as UNIX, Linux, Windows 2000, NT, XP, etc.

Kerberos is used as an authentication protocol in cases where attackers monitor network traffic to intercept passwords. It secures communication, provides single sign on and mutual authentication, and does not send a user's password in the clear on an insecure network.

An alternative solution suitable for digital libraries (Winslett et al., 1997), is to represent information about an individual using credentials. Credentials are "abstract objects which contain statements expressing knowledge or information from a definite context." Credentials do not specify direct information about a client and their attributes, they describe the local environment and context in which the requests originate (Ching et al., 1996).

Digital credentials can be used as a means of authentication in providing DL access control (Winslett et al., 1997). Two agents can be used to assist in the management: a personal security assistant and a server security assistant, to manage digital credentials using a client/server model. The server must notify the client of the credentials required for the current request. The client then sends its credentials for authentication. The client must have some trust of the server to give its credentials, which raises privacy issues.

The personal security assistant is used to obtain credentials on behalf of the client, store the credentials, parse and interpret the required credentials, and manage the acceptance policies (Winslett et al., 1997). A server security assistant is available to specify the credential acceptance policies and their usage.

There is a tradeoff between flexibility and security that must be considered when choosing an access control model, as is discussed below.

2.2.1 Access matrix model

This conceptual model specifies the rights that each subject possesses for each object (Tolone et al., 2005). Actions on objects are allowed or denied based on the access rights specified. There are 2 implementations of the AMM:

- An Access Control List provides a direct mapping of each object the subjects are allowed to access, and their usage rights (owner, read, or write).

- A Capability List defines the objects each subject is allowed to access and the usage rights.

Access control lists and capability lists are not suitable for distributed systems. Their limitations lead to multiple problems (Nagaraj, 2001). ACL provides limited expressibility of policies. Any change in the policies will propagate in the system/application. Authentication in a system that uses ACL solely is a problem because using username & password in a distributed system is not practical. In a distributed system, administration of the system should be decentralized by delegation to reduce the overhead. The owner of the object specifies a policy in ACL. If an overall policy is specified by an entity higher than the object owner, then conflicts may occur in the access rights. The number of administrative entities in a distributed system can be very large. Not all the administrators may have trust amongst themselves, resulting in incorrectly defined policies. For example, admin A may trust B but not C, however B may trust C. If A were to define policy for B then it would be implicitly applicable to C, causing problems.

2.2.2 Role-based access control

Role-based access control involves policies that regulate information access based on the activities the users perform. Such policies require the definition of roles in the system: "a set of actions and responsibilities associated with a particular working activity" (Sandhu & Samarati, 1994). Permissions are assigned to roles instead of individual users. Specifying user authorization involves 2 steps: first assigning the user to a role, second defining the access control that the role has over certain objects.

RBAC is easier to manage and is more extensible than ACL. However RBAC doesn't flexibly handle constraints, where a user with a specific role may need specific permission on an object. An example of RBAC architecture addressing key limitations is OASIS (Bacon et al., 2003), for use in distributed systems. Role management in OASIS is decentralized and service specific. OASIS is integrated with an event-based middleware that notifies applications of any environmental changes. Roles are parameterized by applications and services to define their client roles, and to enforce policies for role activation and service invocation within each session. Role membership certificates (RMC) are returned to each user on successful login, to be used as a credential to activate other roles (Bacon et al., 2003).

RBAC is suitable for use with digital libraries because it supports decentralized architectures and varying roles, however RBAC doesn't allow for the definition of different roles in a collaborative group.

2.2.3 Task based access control

The Task based access control model extends subject/object access control by allowing the definition of domains by task-based contextual information (Tolone et al., 2005). Steps required to perform the task are used to define access control; the steps are associated with a protection state containing a set of permissions for each state, which change according to the task. TBAC uses dynamic management of permissions.

TBAC systems are limited to defining contexts in relation to activities, tasks, or workflow progress. Since it is implemented by recording usage and validity of permissions, therefore, TBAC requires a central access control module to manage permissions activation and deactivation in a just-in-time fashion.

2.2.4 Team based access control

RBAC doesn't address cases where group members of different roles want to collaborate in a single group. The TMAC model defines collaboration by user context and object context. "User context provides a way of identifying specific users playing a role on a team at any given moment" (Tolone et al., 2005) while object context defines the objects required.

TMAC offers the advantages of RBAC along with ability to specify fine-grained control on users and on object instances. A scalable access control data structure can be used with large collections, applying concepts of team based access control, focusing mainly on the access control data structure, and employing an access control framework called Document Access Control Method (DACM) with a Document Storage System (DocSS) (Gladney, 1997). DACM allows the decentralized administration of privileges, the definition of different rule sets to control a single collection, and different delegation patterns as models.

Current object access control policies use an array of rules to record the privileges each subject is allowed to each object. This is impractical to manage in the large data collections found in digital libraries. DACM solves this problem by finding symmetries in a permission function to allow a brief expression without losing important distinctions.

2.2.5 Content based access control

Another approach to access control models involves defining models according to content. This approach is applicable in digital libraries and distributed systems (Adam et al., 2002), where the access rights to the user are dynamic and may change with each login. Content based access control policies are very well suited for digital libraries and distributed systems. Recent research has proposed different models; most use digital credentials for authentication, but vary in the definition/storage of the policy.

An important content based access control model (Ferrari et al., 2002), introduces a content- based access control system, Digital Library Authorization System, that utilizes the Digital Library Authorization Model (DLAM). Subject, object, and privilege sets can't be used to define policies in digital libraries mainly because DLs are dynamic with large collections of data and subjects. It defines access control policies based on subject qualifications and characteristics. DLAM provides a means to specify the qualifications and characteristics of subjects. It uses content dependant and independent access control and allows the definition of policies with varied granularity.

Functionality

The concept of functionality encompasses the services that a Digital Library offers to its users (Gonçalves et al., 2008). The minimum functions of a Digital Library include adding new objects to the library or searching and browsing the library and other functions relating to DL management. A security attack that can affect the functionality of the Digital Library is a Denial of Service attack, which can affect the performance of the system and prevent users from accessing the system.

Architecture

Digital libraries are complex forms of information systems, interoperable across different libraries and so require an architectural framework mapping content and functionality onto software and hardware components (Candela et al., 2007). There are various models for architecture, e.g., client-server, peer-to-peer, and distributed. All these require the protection of the communication channels between 2 parties, where sensitive data might be transferred (Kohl et al., 1998). Securing the connections involves different layers - Internet, transport, or application layer - depending on the architecture of the system.

The distributed model is scalable and flexible. It is useful when building a digital library with changing content from different sources and offers potential for increased reliability. The security requirements for a distributed digital library are challenging, since the content and operations are decentralized. Fault tolerance and error recovery are issues that affect a distributed system. Replication is used to increase the availability of the system. While this approach solves problems with denial of service attacks, it complicates the protection of the content because a replica of the content exists.

The client-server model doesn't have the same security problems as a general distributed model, however, it presents a major security weakness, the server being a single point of failure. Attacks are concentrated on one server rather than on the multiple replicas of a distributed model.

Quality

The content and behavior of a Digital Library is characterized and evaluated by quality parameters. Quality is (Gonçalves et al., 2007) a concept not only used to classify functionality and content, but also used with objects and services. Some of the parameters

are automatically measured and are objective while others are considered subjective; some are measured through user evaluations.

Policy

Policy is the concept that represents the different regulations and conditions that govern the interaction between the Digital Library and users. Policy supports both extrinsic and intrinsic interactions (Candela et al., 2007) and their definition and modification.

Examples of security issues relating to policies include providing digital rights management, privacy, and confidentiality of the content and users, defining user behavior, and collection delivery.

SUMMARY

Digital libraries should be secure. This is an important quality that affects all aspects, as has been shown above using the DL characterization of the DELOS Reference Model (Candela et al., 2007). We also can summarize and elaborate upon this point using another framework for DLs (Goncalves et al., 2004).

The 5S framework supports Societies and their needs, covering all aspects mentioned above about Users and related Policies, as well as Quality (Gonçalves et al., 2007). Since Societies cover software actors, agents, components, modules, etc., this also encompasses related Architectural issues. Thus, security with regard to Societies covers issues like client/server, commerce, identity, peer-to-peer, privacy, rights, roles, teams, and trust.

Scenarios cover functions, operations, requirements, services, and tasks. Examples (Gonçalves et al., 2008) include access, access control, authentication, browsing, copying, denial of service attacks, encryption, payment, recovery, searching, usage, and watermarking.

Spaces cover distributed aspects, as well as representations related to 1D, 2D, 3D, and higher dimensional spaces. These include feature, measure, metric, probability, vector, and topological spaces – used throughout computer and human systems.

Structures cover all types of organization, including data structures and databases, with lists (e.g., access control or capability), graphs, and networks. Structures are overlaid on other constructs in the 5S framework, especially on Streams. Thus, documents are structured streams, while protocols involve scenarios applied to structured communication streams. Structures and Streams cover all types of content, and the many security issues related, including digital rights management, fingerprints, and watermarks.

Clearly, DL security support can be complicated, but the above discussion should help readers organize their thinking and make sure that DL systems meet security requirements.

REFERENCES

1. Adam, N. R., Atluri, V., Bertino, E. & Ferrari, E. (2002). "A Content-Based Authorization Model for Digital Libraries." *IEEE Transactions on Knowledge and Data Engineering* 14(2) : 296-315.
2. Athanasopoulos, G., Fox, E., Ioannidis, Y., Kakaletris, G., Manola, N., Meghini, C., Rauber, & Soergel, D. (2010). A Functionality Perspective on Digital Library Interoperability. *Research and Advanced Technology for Digital Libraries, Proc. 14th European Conference, ECDL2010, Sept. 6-10.* Glasgow: 405-408
3. Bacon, J., Moody, K. & Yao, W. (2003). "Access control and trust in the use of widely distributed services." *Software Practice & Experience (Middleware)* 33(4): 375 - 394.
4. Candela, L., Castelli, D., Ferro, N., Ioannidis, Y., Koutrika, G., Meghini, C., Pagano, P., Ross, S., Soergel, D., Agosti, M., Dobreva, M., Katifori, V. & Schuldt, H. (2007). The DELOS Digital Library Reference Model. http://www.delos.info/index.php?option=com_content&task=view&id=345
5. Ching, N., Jones, V. & Winslett, M. (1996). Authorization in the Digital Library: Secure Access to Services across Enterprise Boundaries. *Third International Forum on Research and Technology Advances in Digital Librarie.* Washington, DC, IEEE: 110 - 119
6. Chowdhury, G. & Chowdhury, S. (2003). *Introduction to Digital Libraries*, Facet Publishing.
7. Dai, W. (2009). "Speed Comparison of Popular Crypto Algorithms." Accessed in 2010, from http://www.cryptopp.com/benchmarks.html.
8. Ferrari, E., Adam, N. R., Atluri, V., Bertino, E. & Capuozzo, U. (2002). "An Authorization System for Digital Libraries." *The VLDB Journal* 11(1): 58 - 67.
9. Fetscherin, M. & Schmid, M. (2003). Comparing the Usage of Digital Rights Management Systems in the music, film, and print industry. *Proceedings of the 5th International Conference on Electronic Commerce.* Pittsburgh, Pennsylvania, ACM
10. Gladney, H. M. (1997). "Access Control for Large Collections." *ACM Transactions on Information Systems (TOIS)* 15(2): 154 - 194.

11. Gonçalves, M. A., Fox, E. A., & Watson, L. T. (2008). "Towards a Digital Library Theory: A Formal Digital Library Ontology." *Int. J. Digital Libraries* 8(2): 91-114
12. Goncalves, M., Fox, E., Watson, L. & Kipp, N. (2004). "Streams, Structures, Spaces, Scenarios, Societies (5S): A Formal Model for Digital Libraries." *ACM Transactions on Information Systems (TOIS)* 22(2): 270 - 312.
13. Gonçalves, M. A., Moreira, B. L., Fox, E. A., & Watson, L. T. (2007). ""What is a good digital library?" - A quality model for digital libraries." *Information Processing and Management* 43(5): 1416-1437
14. Johnson, N. F. & Jajodia, S. (1998). " Exploring Steganography: Seeing the Unseen." *IEEE Computer* 31(2): 26-34.
15. Kohl, U., Lotspiech, J. & Nusser, S. (1998). Security for the Digital Library - Protecting Documents Rather than Channels. *Ninth Workshop on Database and Expert Systems.* Vienna, Austria: 316 - 321
16. Mintzer, F., Lotspiech, J. & Morimoto, N. (1997) "Safeguarding Digital Library Contents and Users." *D-Lib Magazine* 3(7/8), July/August 1997, http://www.dlib.org/dlib/december97/ibm/12lotspiech.html.
17. Nadeem, A. & Javed, M. Y. (2005). A Performance Comparison of Data Encryption Algorithms. *1st International Conference on Information and Communication Technologies.* Karachi, Pakistan, IEEE: 84 - 89
18. Nagaraj, S. V. (2001). Access Control in Distributed Object Systems: Problems with Access Control Lists. *Tenth IEEE International Workshops on Enabling Technologies: Infrastructure for Collaborative Enterprises.* Cambridge, MA, IEEE: 163 – 164
19. Neuman, C. & Ts'o, T. (1994). Kerberos: An Authentication Service for Computer Networks. *IEEE Communications Magazine*, IEEE. 32: 33 - 38
20. Sandhu, R. S. & Samarati, P. (1994). Access Control: Principle and Practice. *IEEE Communications Magazine*, IEEE. 32: 40 - 48
21. Schonberg, D. & Kirovski, D. (2004). Fingerprinting and Forensic Analysis of Mutimedia. *Media Management.* New York, USA, ACM
22. Stallings, W. (2006). *Cryptography and Network Security*, Pearson Prentice Hall.

23. Tolone, W., Ahn, G.-J., Pai, T. & Hong, S.-P. (2005). "Access Control in Collaborative Systems." *ACM Computing Surveys* 37(1): 29 - 41.

24. Tyrväinen, P. (2005) "Concepts and a Design for Fair Use and Privacy in DRM." *D-Lib Magazine* 11(2), February 2005, http://www.dlib.org/dlib/february05/tyrvainen/02tyrvainen.html.

25. Winslett, M., Ching, N., Jones, V. & Slepchin, I. (1997). Assuring Security and Privacy for Digital Library Transactions on the Web: Client and Server Security Policies. *IEEE International Forum on Research and Technology Advances in Digital Libraries (ADL)*. Washington, DC, IEEE: 140 - 151

10
CHAPTER

LIBRARY VALUE IN THE CLASSROOM: ASSESSING STUDENT LEARNING OUTCOMES FROM INSTRUCTION AND COLLECTIONS

Denise Pan[a], **Ignacio J. Ferrer-Vinent**[b], **Margret Bruehl**[c]

[a] *Technical Services, University of Colorado Denver, Auraria Library, 1100 Lawrence Street, Denver, CO 80204, USA*
[b] *University of Colorado Denver, Auraria Library, 1100 Lawrence Street, Denver, CO 80204, USA*
[c] *University of Colorado Denver, Department of Chemistry, Campus Box 194, P.O. Box 173364, Denver, CO 80217-3364, USA*

ABSTRACT

What is the value of library services and resources in the college classroom? How do library instruction and collections contribute to academic teaching and learning outcomes? A chemistry instructor, instruction librarian, and technical services librarian collaborated to answer these questions by combining chemistry education and information literacy pedagogy to assess student learning. The authors developed curriculum units that teach information literacy skills and scientific literature research in a General Chemistry Laboratory course for Honors students. Their study extends beyond examining library instruction and collections assessment in isolation. Rather, their research protocol intends to contribute to student learning outcomes assessment research. The authors propose that an embedded, mixed-methodology, and longitudinal approach can be used to collect data and assess outcomes in terms that describe and measure the value of library services and resources.

Keywords: Value of library services and resources; Assessment; Information literacy; Student learning outcomes; Library instruction; Collections assessment

INTRODUCTION

The institutional goal of research universities should be a balanced system in which each scholar – faculty member or student – learns in a campus environment that nurtures exploration and creativity on the part of every member (Boyer Commission on Educating Undergraduates in the Research University, 1998, p. 10).

This story began with a simple email and lead to an interdisciplinary study conducted over three academic years. At the University of Colorado (CU) Denver, in the summer of 2010, a technical services librarian contacted a chemistry instructor and asked: "In the current economic climate, when university administrators are looking for ways to balance the budget, it is imperative that libraries provide evidence of value and demonstrate their contribution to university priorities. Are you available to discuss your potential participation in a study?" By fall term, the survey participant became a co-researcher and with the help of an instruction librarian, the library was embedded into two chemistry classes. Together the three like-minded faculty members from different areas of academia and librarianship collaborated to produce curriculum units for Honors students in General Chemistry Laboratory I and II courses, administered over fall and spring semesters. The authors exposed first-year students to scientific literature and assigned information literacy activities that help build problem-solving and critical thinking skills to engage and promote student success. Moreover, they developed and implemented a research protocol that enabled them to gather and analyze data on student learning outcomes over time.

At the conclusion of a three-year study, the three faculty members asked themselves three questions: Can we generalize this methodology? Will it scale? Does it contribute to the organizational goals of student retention and success? Their answer was, "we believe so." This article is an invitation to practicing librarians and Library and Information Science researchers to implement the CU Denver research protocol for gathering and analyzing data to measure the value of library services and resources. In this article, the authors will explain how their article contributes to the growing body of literature focused on student learning outcomes assessment; describe their research protocol and curriculum units; and provide a summary of study results. Companion articles address the case study methodology, implementation, and student performance assessments (Ferrer-Vinent, Bruehl, Pan, & Jones, submitted for publication); and describe the curriculum units developed and their connection to building information literacy (Bruehl, Pan, & Ferrer-Vinent, submitted for publication).

Student learning outcomes assessment

Like a rally call to the troops, the Association of College and Research Libraries (ACRL) and Megan Oakleaf created *The value of academic libraries: A comprehensive research review and report* (2010). In essence, the report encourages campus level conversation on assessment, accountability, and value. Within the context of institutional mission and outcomes, they identify "Student Success" as one of the top ten areas of library value on which to focus a research agenda. In response to the question – "How does the library contribute to student learning?" – Oakleaf states that the current literature on information literacy is "voluminous," but a majority is "sporadic, disconnected, and reveals limited snapshots of the impact of academic libraries on learning." Instead, she recommends that "Academic librarians require systematic, coherent, and connected evidence to establish the role of libraries in student learning" (ACRL, 2010, p. 118). In her review and analysis of the literature, Oakleaf introduces several practical suggestions (p. 37–42). The authors distill these concepts into three words that describe the essence of their foray into student learning outcomes assessment: collaboration, purposefulness, and longevity.

Collaboration

Comprehensive and meaningful assessment of student success is impossible in isolation. A learning ecosystem can be cultivated between student and instructor; student and librarian; and instructor and librarian. Poetically described by the Boyer Commission report, *Reinventing undergraduate education*, "The ecology of the university depends on a deep and abiding understanding that inquiry, investigation, and discovery are the heart of the enterprise, whether in funded research projects or in undergraduate classrooms or graduate apprenticeships. Everyone at a university should be a discoverer, a learner" (Boyer Commission on Educating Undergraduates in the Research University, 1998, p. 9). Throughout library literature, these sentiments have been echoed for more than a decade.

When evaluating the 21st century library, Smith (2001) describes the changing environment of higher education. The concept of learning has shifted from the "teacher's knowledge to the student's understanding and capabilities...it requires the faculty to bring the strength of the research paradigm into the learning process" (2001, p. 29). Academic faculty members are being asked to become learning experts by focusing on outcomes assessment — developing individual students' competencies and demonstrating collective programmatic success. As part of the academic community, the mission of the library must change from "a content view (books, subject knowledge) to a competency view (what students will be able to do)" (p. 32). No longer gatekeepers to materials or tools, academic librarians must take a more active role in the learning process and contribute student learning outcomes for academic programs across the curriculum.

Similarly, Nimon recognizes the expanded role of librarians in measuring the outcomes of academic programs. To do so, she encourages developing partnerships between the library and academic departments to teach information literacy. Moreover, the success is contingent on including assessment criteria that reflect the goals of all stakeholders — librarian, academics, and students. She explains, "Student evaluation of the program must be appropriately tailored to show whether its goals were readily visible to the learners and whether the learners considered them met...It will be necessary for the assessment of student work to be at least in part a joint responsibility" (Nimon, 2001, p. 50).

Participation in pedagogy and assessment activities is not just the role of academic librarians, but their obligation. Bundy asserts that they "can no longer responsibly disengage from why students want the print and digital information and resources to which libraries can now so readily provide access. Nor can they disengage from whether those students have the capacity to apply that information well, and to what use they put it" (2004, p. 2). Instead, both academic teachers and librarians should be immersed in the total educational process — including program and curriculum development, learning design, pedagogies, assessment, and the scholarship of teaching and learning.

Information literacy is an important concept that should be owned by all educators (Bundy, 2004, p. 7). However, this terminology may not be recognizable outside the library walls. Some examples of synonyms and overlapping concepts that may resonate more widely include the following: information, research, or 21st century skills; independent scholarship or research; lifelong learning; scientific method; research processes; and Bloom's taxonomy. As Oakleaf explains, "For those facing greater challenges, establishing and using a common language that emphasizes shared campuswide values may produce greater success" (2011, p. 65).

In his model for academic libraries, Lewis (2007) proposes a strategy for maintaining the central library position on campus in the digital age. One of his top five strategies is to "Reposition library and information tools, resources, and expertise so it is embedded into the teaching, learning, and research enterprise...Emphasis should be placed on external, not library-centered, structures and systems" (2007, p. 3). By focusing on student learning, academic libraries and librarians have new opportunities to reestablish their place on campus, engage with their colleagues, and maximize their contribution to their institutions and higher education as a whole.

Purposefulness

When developing student learning outcomes assessments, academic librarians, and faculty members should proceed with purpose. In other words, they are intentionally gathering and analyzing particular types of data. Specifically, Dugan & Hernon (2002) assert, "student learning outcomes are concerned with attributes and abilities, both cognitive and affective, which reflect how the student experiences at the institution supported their development as individuals" (2002, p. 377). In addition, "outcomes assessment alerts us to what students know or do not know about research," Carter explains, "thus allowing librarians to adapt instruction to the needs of the students" (2002, p. 41). This requires a commitment to document, evaluate, and communicate impacts on student learning, as well as improving one's own teaching and assessment skills (Oakleaf, 2011, p. 70).

The literature on information literacy assessment identifies numerous techniques and tools. These tactics require a range of resources or experiences and assess various perspectives. Radcliff et al. describe three different learning domains: *affective* assesses how students feel or their opinions; *behavioral* evaluates what students can do; and *cognitive* measures what students know. They classify performance assessments, such as report writing, as part of the behavioral domain (2007, p. 19–20, 115). In contrast Oakleaf describes performance assessments as "real-life applications of knowledge and skills" and "[they] reinforce the concept that what students learn in class should be usable outside the classroom" (2008, p. 239). All can agree that assignments with an information literacy component can be used to measure higher-order thinking skills, and to achieve greater integration and contextualization in an academic course. While the results can offer a high degree of validity, they may have limited generalizability (Radcliff et al., 2007 and Oakleaf, 2008).

The literature on assessment of information literacy is quite bountiful — well documented in monographs, handbooks, manuals, guides, and articles. The focus is nearly exclusively on library instruction, and overlooks other areas of librarianship. Library collections are included in the broader context of library assessment and emphasize cost effectiveness (Hufford, 2013, p. 20–26). Kinman's article "E-metrics and library assessment in action" is a rare example that highlights the significant role electronic resources play in demonstrating the value of libraries and impacting on student learning outcomes (2009). There is ample opportunity to expand this area of the scholarship into a new line of inquiry that captures the complexity of the learning environment and inspires more rigorous and critical investigation.

Longevity

More than library instruction, information literacy extends outside the confines of the library and extends to the classroom, campus, and beyond. Each level requires more collaboration and complex logistics (Iannuzzi, 1999, p. 304). To be truly meaningful, outcomes assessment is an ongoing process. It requires time, effort, resources, and collaboration with academic faculty. The research process can be simple or elaborate, but must be based on meaningful data. From each experience, researchers can learn, refine, and improve assessment procedures (Carter, 2002, p. 41). Oakleaf & Kaske (2009) recommend that librarians follow best practices, use multiple methods and strategies, adjust goals and objectives, and repeat assessments continuously over time (2009, p. 283).

CU Denver research protocol & curriculum units

The literature describes collaboration, purposefulness, and longevity as key ingredients to achieve student learning outcomes assessment. This article continues this conversation. The authors at CU Denver adopted these assessment perspectives and adapted them to their local circumstances. Specifically the authors summarize their student learning outcomes assessment research protocol in three components: collaborating internally and externally; articulating outcomes; and assessing over time. By doing so they are able to expand beyond their silos and approach student learning outcomes assessment from an interdisciplinary perspective. The results of their study demonstrate the rich data and insights that are possible from their approach.

Collaborating internally and externally

The CU Denver researchers identified multiple opportunities for collaborations within librarianship, academia, and the campus. They found commonality with academic faculty and librarians from diverse areas and expertise. In addition, they also created teaching, research, and learning opportunities between faculty and students. A visual representation of these connections is available in Fig. 1.

At CU Denver, librarians have faculty status. The library and chemistry departments are two separately administered academic units reporting to the Office of the Provost. Although the departments often cooperate on collections and instruction initiatives, the authors' research study represents a new level of collaboration. Library instruction and collections assessment activities are embedded into two Chemistry courses. More specifically, the researchers focused on shared values and appreciated other perspectives from different disciplines. In essence, they are proponents of a "golden triangle" approach to curriculum design "where everyone contributed to each component with the outcome being more robust than if they used the more linear model" (Fox & Doherty, 2012, p. 151–2).

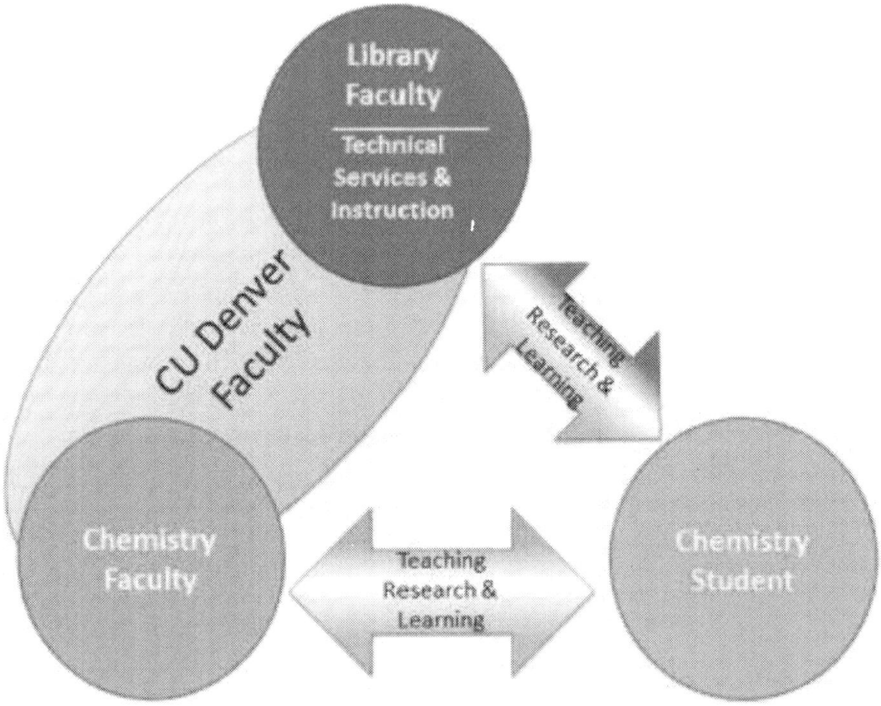

Figure 1. Research relationships.

Within the library context, a technical services librarian and an instruction librarian from the same academic library developed a partnership to conduct a student learning outcomes assessment research study with a chemistry instructor. This experience may be atypical. Bundy states that "few university libraries or librarians directly engage with, or reach out to, other parts of the profession...the focus tends to be on information resource sharing and access, rather than on learning collaborations and strategies" (Bundy, 2004, p. 9).

Most significantly, the CU Denver faculty developed a research relationship with their students. With Colorado Multiple Institutional Review Board approvals, they were granted permission to gather data from the students. In her Post Laboratory Assignment, Bruehl specifically invited her students to participate in the study and assured them that no additional work would be assigned, "Simply completing this assignment is all that is required." All personal identification was removed by Bruehl before Pan and Ferrer-Vinent analyzed the student data. Bruehl also explicitly told the students that they could contact her during lab or via email to option out of the study and their data would be excluded from the study (Bruehl et al., submitted for publication).

Articulating outcomes

Sharing common values and goals, the authors found exciting synergies for developing student learning outcomes assessment. All three researchers aspired to develop a project that could provide compelling evidence of supporting the institution's mission for campus administrators. They surmised that an assignment that concentrated on information literacy could help them achieve this objective. In doing so, they developed a research instrument that enabled them to gather data on students' ability to find, access, and evaluate scientific literature.

When developing her Honors General Chemistry I and II Laboratory courses, Bruehl aims to create a student centered learning environment. This teaching philosophy encourages active participation by students while the learning process is facilitated by the instructor. The investigative and inquiry-based laboratory courses emphasize teaching students problem-solving and critical thinking skills through open-ended experiments and using specialized techniques and instrumentation. This teaching and learning strategy builds on the CREATE (Consider, Read, Elucidate hypothesis, Analyze and interpret data, Think of the next Experiment) method.

Typically, the real language and research process documented in primary scientific literature is not presented to students until their upper division course or not at all during their undergraduate education. This delay is recognized as a missed opportunity at best, or a loss of science majors at worst. Using the CREATE pedagogy exposes students to scientific research literature and demonstrates the creative and exploratory nature of collecting and interpreting data for research. By doing so, students are being taught how to think like scientists, increase engagement, and could help with retention (Hoskins et al., 2007 and Gottesman and Hoskins, 2013). Findings from one study indicate that the CREATE method "increases students' confidence in their ability to read and understand primary literature, improves their self-assessed understanding of the nature and processes of science, and encourages their development of more sophisticated epistemological beliefs" (Hoskins, Lopatto, & Stevens, 2011, p. 375).

The CU Denver research study has two overarching goals. The first is to demonstrate that teaching information literacy skills in first-year chemistry courses can provide immediate and long-term benefits to student performance. The second is to quantify the benefits students received in their educational activities by using library collections. Long term, the authors hope to be able to provide campus administrators with evidence that their efforts contribute to meaningful outcomes, such as student retention, graduation rates, and economic benefit.

To achieve these goals, the authors developed two complimentary curriculum modules to establish a foundation of information literacy skills for beginning science students using scientific literature and laboratory experiments. The first unit, administered in the fall semester, is entitled "*Introduction to Scientific Literature*" and is comprised of three components: 1) formal library instruction by a science librarian; 2) reading and in-class discussion on a scientific journal article; and 3) pre- and post-lab exercises to research an idea for a new general chemistry experiment. The students are asked to develop information literacy skills by searching the scientific literature for concepts, recording their research process, refining their premise, writing a brief description of their proposed experiment, and citing three resources in American Chemical Society (ACS) format that directly support their proposal.

Building on the skills and information they learned in the fall, in the spring semester they are asked to conduct a multi-week project entitled "*Design your own General Chemistry Lab.*" Again students explore the scientific literature for ideas on which to base their experiment, and record the reference for any resources they downloaded and read. Their final project includes a formal proposal; documented laboratory procedure that has been designed, developed, and tested by the student; citation list in ACS format; and presentation of results to the class.

In both semesters, students are conducting research in the scientific literature. To record their research process, Bruehl requires students to complete a research process template. The format of the template is a table with three separate columns to record: 1) database/search tool name; 2) search terms/refinements; and 3) resources viewed. The third column instructs the students to "paste ACS or specific journal format citation here." Table 1 includes an example of a completed research process template, with good searching and narrowing values, provided by a student in spring 2012.

Table 1. Sample research process template.

Database/search tool name	Search terms/refinements	Resources viewed (paste ACS or specific journal format citation here)
Journal of Chemical Education	Simple and titration and experiment	Lopez, E.; Vassos, B.H.. Colorimetric titration experiment for the undergraduate laboratory. *J. Chem. Educ.* **1984**, *61(11)*, 1025.
Journal of Chemical Education	Titration and content of	Deal, S.T.; Pope, S.T.. Improved end point detection in the redox titration of vitamin C in green peppers. *J. Chem. Educ.* **1996**, *73(6)*, 547.
Journal of Chemical Education	Titration and vitamin content	Sowa, S.; Kondo, A.E., Sailing on the "C": A vitamin titration with a twist. *J. Chem. Educ.* **2003**, *80(5)*, 550.
Journal of Chemical Education	Titration lab and vitamin C	Vitz, E.. A student laboratory experiment based on the vitamin C clock reaction. *J. Chem. Educ.* **2007**, *84(7)*, 1156.
Journal of Chemical Education	Titration lab and vitamin C	Conn, L.W.; Johnson, A.H., Vitamin C. Content of frozen orange and grapefruit juices. *J. Chem. Educ.* **1933**, *25(2)*, 218–221.
ScienceDirect	Titration and vitamin C	Suntornsuk, L.; Gritsanapun, W.; Nilkamhank, S.; Paochom, A. Quantitation of vitamin C content in herbal juice using direct titration. *J. Pharm. Biomed. Anal.* **2002**, *28(5)*, 849–855.
ScienceDirect	Titration and vitamin C	Lenghor, N.; Jakmunee, J.; Vilen, M.; Sara, R.; Christian, G.D.; Grudpan, K. Sequential injection redox or acid–base titration for determination of ascorbic acid or acetic acid. *Talanta.* **2002**, *58(6)*, 1139–1144.
Google	Vitamin C titration equations	Bacon, J.R. Determination of vitamin C by an iodometric titration. *Western Carolina Univ.* **2006**.

The research process template is an essential assessment tool for the authors to gather data to evaluate the students' ability to find, access, and evaluate scientific literature. Bruehl required the student to record *all* resources viewed, even if the student only read the abstract or the table of contents, and not the article itself. She explains that, "Some resources that you record in this template will be more important to your project. These resources help to formulate your design, and they must ALSO be listed in your formal citation list" (Bruehl et al., submitted for publication).

By documenting the names of database/search tool, search terms, and resources viewed, the authors could see how students are selecting the resources viewed — rejecting or accepting results that fall in and out of the scope, or distract or meet the needs of their topic. Broad subjects are narrowed with keywords, and then filtered down by publications focused on science experiments, such as the *Journal of Chemical Education*. Combining the expertise from academic faculty and faculty librarians, the authors analyze and compare the data from the research process templates and citation lists. Bundy summarizes this partnership and librarian contributions to teaching pedagogy as "Disaggregated roles, such as assessing learning resources for quality, overlap with what librarians do now, and the subject expertise of the academic teacher is being married with the librarian's navigation and sense making of the information universe" (Bundy, 2004, p. 9).

Assessing over time

Realizing that outcomes assessment is a continuous process, the CU Denver researchers intentionally developed the project to be repeatable over multiple semesters and academic years. The process encourages collaborative teaching and learning from the students and one another. The authors refine and improve instructional strategies and assessment methods at each encounter. For example, during citation analysis Ferrer-Vinent and Pan observed that a few students were citing articles that were not available full-text from the library. While it is possible that the students could have requested the article from interlibrary loan, the librarians were worried that these students did not understand the difference between an abstract and the full-text article. They shared their concerns with Bruehl who added this information to her lecture on scientific literature.

Exposing students to information literacy concepts over multiple semesters reinforces skills and the interconnectedness of information. The importance of the research skills and process is reiterated over two terms of an academic year. In the spring semester "*Design your own General Chemistry Lab*" assignment, Bruehl reminds students to expand their understanding beyond their own experience, and to use the research expertise they developed the previous semester. Scientific literature offers "a vast array of resources" to assist them with their assignment, which in turn helps them to develop and recognize sound scientific procedures, evaluate experimental methods, and draw appropriate conclusions. As a result, beginning chemistry students are fine tuning their ability to use the scientific method to investigate a scientific question.

Moreover the CU Denver study was conducted over three academic years. The researchers analyzed consistencies and variations over the same and consecutive semesters across multiple academic years. For example they could examine data for each term (e.g. Fall 2010 and Spring 2011), or comparison between terms (e.g. Fall 2010 versus Spring 2011), and within an academic year (e.g. 2010–2011). Most importantly, since the chemistry students agreed to participate in the research project, the researchers surveyed the same students 1–2 years after they finished the Honors General Chemistry II Laboratory course. Therefore, with the CU Denver research protocol, the researchers gathered quantitative and qualitative data from the students at different points of their academic careers — during their first-year and beyond.

CU Denver research study results

On their own, the researchers could have independently assessed outcomes from their own disciplines' perspectives. By collaborating internally and externally, articulating outcomes, and assessing over time, the authors developed and implemented their research protocol from fall 2010 to spring 2013. Expanding beyond their silos and using an interdisciplinary perspective and mixed methodology approach enabled them to produce distinct student learning outcomes assessment data and analysis.

Prior to their partnership, each author individually engaged in assessment activities. Bruehl conducted student learning outcomes assessment by grading students' assignments, exams, and overall performance in the class. Ferrer-Vinent provided in-class library instruction, surveyed student perceptions, and assessed students' post-instruction performance. With no direct contact with students, Pan measured the library's potential contributions by analyzing usage statistics of electronic resources.

Together they leveraged personal expertise to enhance their shared research project. Their interdisciplinary approach allowed them to develop and evaluate richer data sets. The CU Denver researchers apply the quantitative citation analysis data from the research process template and qualitative surveys to address their research inquiries. Specifically, they intend to demonstrate that teaching information literacy skills in first-year chemistry courses can provide immediate and long-term benefits to student performance; and to quantify the benefits students received in their educational activities by using library collections. Research outcomes are summarized in three subsequent sections: 1) quantitative information literacy study results; 2) quantitative collections ROI study results; and 3) qualitative student benefits survey results. As previously mentioned, details on the authors' research methodology is described in their companion case study article (Ferrer-Vinent et al., submitted for publication).

Quantitative information literacy study results

By recording the references of resources viewed and cited in their assignment, the authors could begin to quantify and aggregate how students are developing information literacy skills. For example, Table 2 demonstrates that they were able to calculate the average number of resources viewed per student to complete the fall and spring term assignments and their average course grade. For the fall semester assignment, students were instructed to use their library skills to view resources and then select three upon which to base a proposed experiment. In spring, students received no guidance on how many resources should be viewed or selected to support their experimental design. Excluding the first semester, during which the authors were still refining the assignment guidelines, students on average looked at almost 10 resources in the course of their assignments and their average grade was 87.5%.

Table 2. Resources viewed and final course grade.

	# of students	Total # resources viewed	Min # resources viewed	Max # resources viewed	Average # resources viewed	Average final grade %
AY2011						
Fall 2010	26	174	3	16	6.7	89.8%
Spring 2011	21	197	1	28	9.4	90.7%
AY2012						
Fall 2011	37	369	3	39	10.0	84.5%
Spring 2012	36	341	0	66	9.5	87.1%
AY2013						
Fall 2012	25	248	5	26	9.9	87.9%
Spring 2013	16	155	4	13	9.7	87.1%
Total	161	1484	0	66	9.2	87.5%

The researchers intentionally collected grades to enable them to correlate academic performance and use of library resources. Said another way, they hypothesize that there is a relationship between the extensive use of library resources and a high score. If students consulted multiple databases, reviewed many articles, and selected the best three resources, they should receive a higher score than a student who only used one database and looked at a few articles. The authors collaborated with Galin Jones – Associate Professor at the University of Minnesota, School of Statistics – to determine that there is a statistically significant positive relationship between resources viewed and the students' final grade for the course. The full explanation on the linear regression analysis is available in the case study companion article (Ferrer-Vinent et al., submitted for publication).

Quantitative collections ROI study results

The authors also use the citation analysis data to calculate the cost benefits of purchasing scientific literature to support learning objectives. To do so, they measure library value by deriving value using return on investment (ROI). According to Tenopir (2012), "Derived values, such as return on investment (ROI), use multiple types of data collected on both the returns (benefits) and the library and user costs (investment) to explain value in monetary terms" (2012, p. 6). In order to measure return on investment, the authors needed to assign a monetary value to having access to non-market resources/services or library collections. They applied contingent valuation as defined by Megan Oakleaf to assign value to library collections and to determine potential willingness to pay to maintain the existence of library collections (ACRL, 2010, p. 50). Many publishers, including ACS, sell individual articles online. If the library did not provide these articles, students could purchase these ACS articles on their own. The authors utilize the price to download an article to derive a market value for the library service of providing access to collections. In turn, this calculated value enabled the researchers to calculate student benefits for the ROI calculations.

The CU Denver Student ROI Model is based on established ROI and cost benefit analysis (CBA) formulas. ROI is calculated as a percentage. It shows the return or increase in value on dollars spent to achieve a benefit. The generic formula is benefits minus costs divided by costs and multiplied by 100.

$$((\text{BENEFITS}-\text{COSTS}) \div \text{COSTS}) \times 100 = \text{ROI}.$$

CBA uses the same values as ROI. However, CBA is the ratio showing the dollar value of benefits gained for dollar value of costs. The basic formula is benefits divided by costs.

$$\text{Benefits} \div \text{Costs} = \text{CBA}.$$

Benefits are the estimated cost for students to buy cited articles directly from the publisher with pay-per-view. Costs are the Library costs to supply cited online journals. This methodology is based on a parallel CU Faculty ROI Model and multi-campus study Pan conducted to measure the institutional value of library resources used by faculty in their research (Pan et al., 2013).

The authors analyzed the citations and determined that students viewed 1489 resources over three years. Most of these articles were searched on the ACS Journal Database 58.0% of the time (863 out of 1489). Collectively the students looked at 1153 journal articles from 200 unique journal titles, and they consulted the *Journal of Chemical Education* (*JCE*) 66.7% of the time (769 out of 1153). Since a majority of the students used the *JCE*, the use and cost estimates for this journal title was used to calculate the ROI and CBA for the course. The authors derived the student costs from the number of times *JCE* was viewed, multiplied by the pay-per-viewed cost. The library cost for *JCE* was estimated by taking the database cost divided by the number of titles in the database.

Since ROI is calculated as a percentage, values over 100% are a positive ROI. Similarly, since CBA is a ratio showing how much is gained for every dollar spent, $1.00 is the breaking point. Values greater than $1.00 are a "positive" CBA. The results of the CU Student ROI Model for Honors General Chemistry I and II Laboratory courses at CU Denver from fall 2010 to spring 2013 are summarized in Table 3. Overall, the results indicate a very strong

ROI and CBA for the *Journal of Chemical Education* with the greatest ROI of 2324% and CBA of $24.24 in fall 2011.

Table 3. Students ROI study results for *Journal of Chemical Education*.

	ROI	CBA
AY 2011		
Fall 2010	626%	$7.26
Spring 2011	884%	$9.84
AY 2012		
Fall 2011	2324%	$24.24
Spring 2012	1109%	$12.09
AY 2013		
Fall 2012	1971%	$20.71
Spring 2013	288%	$3.88[a]
Total	1144%	$12.44

[a] The researchers attribute the low ROI of 288% and CBA of $3.88 in spring 2013 to low student enrollment due to a scheduling conflict and cancellation of one section of the course.

Qualitative student benefits survey results

In addition to collaborating on quantitative citation and cost benefit analysis, the researchers also conducted qualitative assessments. After the students completed the Honors General Chemistry II Laboratory course, Bruehl emailed them an anonymous survey. This was an opportunity to gather data on the impact of teaching information literacy to first-year chemistry students over the long term. The questionnaire reminded the students that "as part of your lab work, you used the [scientific] literature and captured your on-line searches in a research process template," and invited them to participate in the survey as a "follow up to those activities." A copy of the longitudinal student survey is available in the companion article focused on the curriculum units (Bruehl et al., submitted for publication).

The responses to the survey were overwhelmingly positive. Over 40% (37 out of 88) of the students completed the questionnaire. When asked about their experiences before and after Honors General Chemistry Laboratory, more than half stated that they searched a scientific literature database since course completion. In addition, this was nearly a 40% increase over their experience prior to taking the course (see Table 4).

Table 4. Reponses to follow up survey.

	# of students surveyed	# of student responses	# of responses that searched database *prior* to class[a]	% of response *prior*	# of responses that searched database *prior* class[b]	% of responses *since*	% change increase
AY 2011	23	8	4	50%	8	100%	50.0%
AY 2012	40	18	9	50%	17	94%	41.5%
AY 2013	25	11	1	9%	6	55%	37.9%
Total	88	37	14	38%	31	84%	38.6%

[a] Answered "yes" to the question "Prior to your enrollment in Honors General Chemistry Laboratory, had you ever searched a scientific literature database?"
[b] Answered "yes" to the question "Since you have completed Honors General Chemistry Laboratory, have you searched a scientific literature database?"

Surprisingly, students reported that they searched the American Chemical Society (ACS) journals for scientific literature more frequently than Google (77.4% and 71.0% consecutively). SciFinder, ScienceDirect, and Web of Science were also identified as scientific literature databases that they used after finishing the course. The two other resources identified were PubMed and JSTOR (more details provided in Table 5).

Table 5. Database usage since completing Honors General Chemistry Laboratory.

	# of responses that searched database *since* class	Scientific literature databases used						
		ACS	Google	ScienceDirect	SciFinder	Web of Science	Wiley	Other
AY 2011	8	6	6	4	7	5	0	0
AY 2012	17	14	13	12	9	7	3	2
AY 2013	6	4	3	2	3	3	0	0
Total #	31	24	22	18	19	15	3	2
% of Total		77.4%	71.0%	58.1%	61.3%	48.4%	9.7%	6.5%

More or less than half of the students responded that they did the search for their own curiosity or for a research project/internship (see Table 6). The vast majority of students, however, conducted these database searches for another course. The knowledge they acquired was directly applied to 17 other Chemistry courses. More significantly, their information literacy skills transferred to 21 courses in Biological Sciences, 2 in English, and 1 in Psychology. These findings support the authors' hypothesis that teaching information literacy skills can provide students with long-term benefits.

Table 6. Explanation for database searches.

	# of responses that searched database *since* class	Explanation			
		Course	Research project/internship	Curiosity	Other
AY 2011	8	8	3	6	
AY 2012	17	13	9	9	1
AY 2013	6	7	2	1	
Total #	31	28	14	16	1
% of Total		90%	45%	52%	3%

11
CHAPTER

ACADEMIC EDUCATION IN LIBRARY AND INFORMATION MANAGEMENT IN BULGARIA

Rositsa Krasteva[1], Tereza Trencheva[2], Sabina Eftimova[2], Tania Todorova[2]

[1]*Computer Science Department, State University of Library Studies and Information Technologies, Sofia, Bulgaria*
[2]*Library Management Department, State University of Library Studies and Information Technologies, Sofia, Bulgaria*

ABSTRACT

The purpose of this publication is to present the contemporary aspects of training educational and qualification degree "Bachelor" of Specialty "Library and Information Management" of the Library Management Department at the State University of Library Studies and Information Technologies (SULSIT) in Sofia, Bulgaria.In view of specificity and completeness of the presented information in this publication there is a limit, which refers to the training only in educational and qualification degree "Bachelor". The following methods are used: a study of the curriculums of many universities worldwide teaching in this or a related specialty; comparative analysis; synthesis of the obtained information. Accent is put on the disciplines Intellectual Property, Standardization in Library Activities, Quality Management in Library and Information Activities, Library Psychology and Bibliotherapy. The research draws attention to some aspects insufficiently covered in the curriculum of the programs preparing future highly knowledgeable, trained library and information managers, and offers some solutions, based on our experience in the State University of Library Studies and Information Technologies, to the attention of the LIS academic and professional community.

Keywords: Library Management, Information Management, Higher Education, Bachelor's Degree, SULSIT, Bulgaria

1. INTRODUCTION

Socio-economic and political transformations in the Bulgarian reality necessitated a change in the functions of libraries and enhanced their role in the process of accession to the European and World educational, cultural and economic space. Bulgaria is member of the EU since January 1st, 2007. The libraries in accordance with the principle of wide and equal public access to library and information resources, and as a democratic social institutions contribute to social stability, preservation and development of the spiritual and scientific potential of society. The contemporary challenges facing the libraries, suggest the application of adequate library management. It evolves against the background of changing the overall management paradigm—a process characteristic of global management. The outdated management approaches to the concept of "Rational Management" shall be replaced by the Marketing Paradigm, characterized by adaptability and flexibility of management. This undoubtedly imposes new requirements, including management knowledge and skills to the competencies of young specialists, finishing their high degree in specialties like Library and Information Science.

Evaluating LIS Education in Europe, the experts of the European LIS Curriculum Project conclude that in Europe, almost all library and information (LIS) programs in higher education have been developed and are offered within the context of a nation-state [1] . Several authors reported the need for internationalization of library education in Europe in accordance to achieve greater results in student and teacher mobility [2] - [4] . The situation is summarized in LIS Education in Europe: "The Structure and contents of LIS courses vary very much between the different types of LIS education providers in Europe, which include many fairly small academic environments. The apparent disparate nature of LIS educational programmes in Europe constitutes a barrier to increased cooperation in the field. There is a marked need for joint discussions of the structure and contents of LIS school curricula and for identifying and discussing possible common curricular elements both for the purpose of enhancing the quality of individual LIS educational programmes and for the sake of increased collaboration between European LIS school programmes" [5] . Based on these findings, in 2005 started the development of an international project LIS education in Europe, coordinated by the Royal School of Library and Information Science, Denmark. Main goals of the project are:

- To create better opportunities for European mobility of students and teachers;
- To increase the scale of mobility and inter-institutional cooperation;
- To develop a common conceptual framework to define the core of mandatory elements within the LIS curriculum as a basis for increasing the mobility of students and teachers and the acceleration of the Bologna process;
- To work for greater flexibility, transparency and comparability of curricula.

The results of this project were published in the European Curriculum Reflections on Library and Information Science Education [6] . In a separate section was provided information related to the training of library management. Focus is on disciplines such as Marketing Information Services; Communication Skills and Negotiations; Intellectual Property and Information Law; Quality Management and etc.

Taking into account the achievements of our colleagues in Europe, in this article we would like to share our experiences, best practices and suggestions on training in the undergraduate program "Library and Information Management" at the State University of Library Studies and Information Technologies in Sofia, Bulgaria. In response to the social and professional needs, in 2008 in the structure of the Faculty of Librarianship and Cultural Heritage (FLCH) in SULSIT is created new Department of Library Management. Founder and first head of the department was Prof. DSc Ivanka Yankova (from 2011 she is a Dean of FLCH). Since June 2011 up to now— Assoc. Prof. PhD. Tania Todorova is a Chair Holder of Library Management Department. The main task of the Library Management Department is the preparation of bachelors, masters and doctoral students in the field of library and information management.

2. GOAL, RESEARCH TASKS AND METHODOLOGY OF THE STUDY

The purpose of the study is to present the contemporary aspects of training in educational and qualification degree "Bachelor" of the Specialty "Library and Information Management" of the Library Management Department at the State University of Library Studies and Information Technologies (SULSIT) in Sofia, Bulgaria and to make comparison with the curricula of some of the biggest universities worldwide, which are accredited in LIS education. In view of specificity and completeness of the presented information in this publication there is a limit, which refers to the training only in educational and qualification degree "Bachelor". To achieve the objective are set out the following major research tasks, outlining the methodology of the study: 1) Presentation of the Library Management Department in SULSIT (Bulgaria) and its activity; 2) Exploration, presentation and comparative analysis of the curriculums of many universities worldwide teaching in this or a related specialty; 3) Conclusions and recommendations. The methodology for achieving the objective of the study and solving the set research tasks include the following specific methods: method of study and content analysis, comparative analysis; synthesis of the obtained information. Accent is put on the disciplines Intellectual Property, Standardization in Library Activities, Quality Management in Library and Information Activities, Library Psychology and Bibliotherapy. The research draws attention to some aspects insufficiently covered in the curriculum of the programs preparing future highly knowledgeable, trained library and information managers, and offers some solutions, based on our experience in the State University of Library Studies and Information Technologies, to the attention of the LIS academic and professional community.

3. INTERNATIONAL COOPERATION OF THE LIBRARY MANAGEMENT DEPARTMENT

The creation and development of the Library Management Department and proposed programs—Bachelor Program "Library and Information Management" and Master Program "Library, Information and Cultural Management" are influenced by the principles and guidelines of the Bologna Declaration and the study and application of leading experience of European and foreign universities.

In the period 2009-2013, the Library Management Department was a partner in two Erasmus Intensive Programmes with the participation of European universities. The Project Das Grimm-Zentrum-(k)ein Bibliotheksmärchen) (2009-2011) with coordinator Humboldt University in Berlin (Germany) created a unique opportunity for cooperation with University of Vienna (Austria), Vilnius University (Lithuania), Masaryk University Brno (Czech Republic) [7] . In the period 2011-2013 Library Management Department at SULSIT was coordinator of Erasmus Intensive Programme "Library, Information and Cultural Heritage Management—Academic Summer School" [8] . Project Manager is Tania Todorova—Head of Library Management Department. Our partners are: Hacettepe University in Ankara (Turkey), University of Zagreb (Croatia), University Paris Descartes (France) and University of Szeget (Hungary). The mission of this project is through using a rich methodological toolkit to implement a modern educational process aimed at implementation of interdisciplinary knowledge and skills relevant to the new requirements in the career development of students in library, information and cultural sector and the policy response to higher education and the EU initiative on "New Skills for New Jobs". The main topics in IP LibCMASSare: Library, Information and Cultural Management; Information Literacy; Preservation and access to Cultural Heritage; Digital libraries; Intellectual Property; Information brokerage; Information technologies in libraries, archives, museums and other cultural institutions [9] [10] .

In the process of creating and updating the curriculum of the Bachelor Program "Library and Information Management" important was the impact of active participation of professors and students from the Library Management Department in the short-term mobility under the Erasmus Intensive Programmes. On Figure 1 is shown the dynamics of the participation of professors. On Figure 2 is presented the students participation. It is necessary to highlight that in 2013 was implemented successfully 1 student mobility of an Erasmus Summer Semester in HochschulefuerMedien in Stuttgard, Germany.

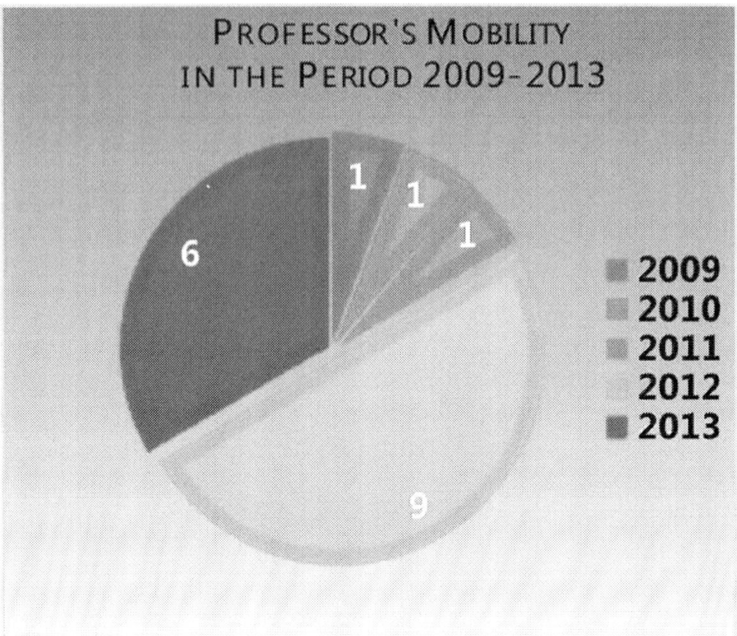

Figure 1. Professor's mobilityfor the period 2009-2013.

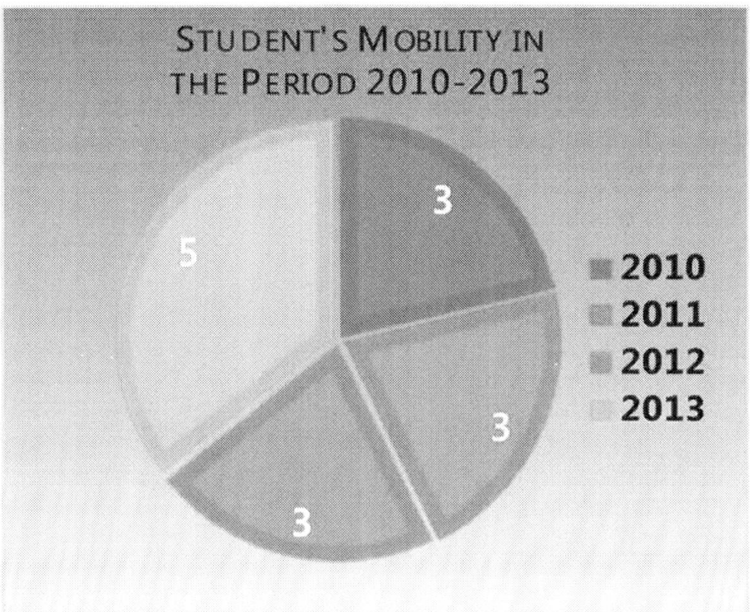

Figure 2. Student's mobilityfor the period 2010-2013.

These projects were unique opportunity for establishment a stable international network in higher education in Library and Information Sciences, Computer Sciences and Cultural Heritage Sciences and to promote cooperation between academic education and practice—library, information and cultural sector.

On 20th November 2011 the Memorandum for establishment of the UNESCO Chair' ICT in Library Studies, Education and Cultural Heritage' was signed between Irina Bokova, the General Director of UNESCO and Prof. DSc Stoyan Denchev, the Rector of the State University of Library Studies and Information Technologies. Many colleagues from Library Management Department collaborate actively in the management and in various activities and projects of the UNESCO Chair [11] .

We could summarize that for five years period from the establishment of Library Management Department in SULSIT we have achieved a presence in the single educational space in the field of Library and Information Sciences and we are highly motivated for future initiatives and partnerships.

4. CURRICULUM AND METHODOLOGY OF LIBRARY AND INFORMATION MANAGEMENT BACHELOR'S PROGRAM

The students which obtained the educational and qualification degree "Bachelor" in "Library and Information Management" in SULSIT must have a thorough scientific, theoretical and practical training in the specialty, which includes:

- Fundamental training on the nature and the diversity of libraries and other cultural institutions. History of the Field;
- Knowledge and understanding of the library cataloguing, acquisition and bibliographic processes and operations;
- Fundamentals of Library Management, Knowledge Management, Standardization and Quality Management in Library and Information Activities;
- Thorough knowledge of modern library and information technologies and their application to all facets of Library and Information Products and Services. Digitalization and Preservation of Library Funds;

- Theoretical and practical training in Information Resource Management, Information Retrieval, Assessing of Information Needs, Intellectual Property and Information Literacy;

- Specialized training in Library, Information and Project Management to acquire knowledge and skills in planning, organizing and controlling the overall operations of libraries, information centers, community centers and other cultural and public institutions;
- Basic knowledge of Marketing, Public Relations and Library Psychology for optimal organization of information and the use of methods to interact with audiences for library collections, services and events.

Experts should be able:

- To raise and resolve the duties arising in connection with the organization and ensure optimal use of information resources in libraries and other organizations with arrays;
- To develop and manage projects;
- To develop and implement programs for digitization and creating their own electronic resources;
- To organize and manage various initiatives to interact the various audiences;
- Carry out self-sociological, psychological and other research needed to analyze the quality of the services and the extent of their approval by the users, and to be able to improve the characteristics of information and communication forms used in the library, and to bring new services or forms;
- To collaborate effectively with national and international LIS community [12] .

Forms of education in educational and qualification degree "Bachelor" are regular and part time and are implemented in eight semesters. Table 1 presents the disciplines currently included in the curriculum for compulsory training of specialty "Library and Information Management" [13] .

Table 1. Compulsory subjects taught in the educational and qualification degree "Bachelor" in specialty "Library and Information Management" in SULSIT.

Discipline	Semester
History of the Cultural Institutions and Librarianship	First semester
History and Theory of Culture	
Information Resources	
Annotating and Referencing of Documents	
Information Systems	Second semester
Book Science	
Rare and Valuable Library Collections	
Bulgarian Literature	
English	
Second Foreign Language	
Sport	
National Bibliography	Third semester
Management and Library Collections Development	
Library Management	
Automated Libraries—Part 1. Cataloguing, Classification and Subject Indexing of Information Recourses.	
Automated Libraries—Part 2. Automation Library and Information Systems. OPAC, Union Catalogues, National and International Cooperation. Application of ICT in Libraries.	
Intellectual Property	Fourth semester
Information Services in Libraries	
Innovations for Preservation of Library Funds	Fifth semester
Library Marketing	
Information Literacy—Programs and Models	
Architecture of the Libraries	
Library Psychology	
Practicumin various types of libraries	
Information Management	Sixth semester
Knowledge Management	
Bibliometry	
Standardization in Library Activities	
Public Relations	
Library Legislation	
Quality Management of Library and Information Activities	Seventh semester
Digitization and Copyright	
Academic Writing	
System Analysis	Eight semester
Project Management	
Practicum and Fieldwork. Development of Bachelor Thesis.	

*Teaching English and sport is foreseen as a compulsory subject for each semester.

There is an option for the students to choose a discipline according to their interests (elective courses). Among the most popular are: Applied Software, Web Design, Content Management Systems, Intellectual Property in Internet, Libraries and Local Authorities and Communities, European Communication Policy. Common culture among students is supplemented by the chosen of them optional subjects such as: Bibliotherapy, Access to information for people with special needs etc. Methodology of teaching the students is done except to traditional methods such as lectures and seminars, and through interactive and situational methods, application of the approach "learning by doing", "edutainment" and etc., which activates the students' participation. Individual forms of work such as term papers, presentations, individual assignments, communication in e-learning platforms—support the learning process and the current verification of students' knowledge. Library Management Department features with the active involvement of students in research projects and joint research papers for participation at scientific conferences and symposiums.

There are two initiatives of the young scientists in the Library Management Department, which are directed to the students. One of them is the work of students in relation to the Week Dedicated to The International Day of the Book and Copyright Day (23 April), with coordinator Dr. Lubomira Parizhkova, where students from all specialties take part into this initiative and have the opportunity to publish their first scientific research in the thematic research collection [14] . The second initiative is led by Dr. Tereza Trencheva and is dedicated to The International Intellectual Property Day (26 April), where the students have the opportunity to meet specialists from practice in the field of Intellectual Property [15] .

For self experience activities the students of Library and Information Management visited: libraries, museums, galleries, workshops for making paper and books, book fairs (with conducting inquiries), book premieres, meetings with authors, public lectures etc. Also, they participate in the initiatives of promotion of reading—marathon of reading, book crossing movement and in different forms of stimulation of children's reading.

Important project, which is creating the opportunity to use the acquired theoretical knowledge in a real working environment is the National Project' Students' Practices' of the Bulgarian Ministry of Education and Science, implemented under the Operational Programme "Human Resources Development" of the EU. For the period July 2013-July 2014, more than 40 students of Library and Information Management Specialty conducted 240 hours internship in libraries, archives and information centers. Our students are actively involved in initiatives of the Bulgarian Library and Information Association and the Association of University Libraries, also. Our goal is nurture an active attitude towards the library profession and knowledge of the real challenges, problems and achievements in their future professional environment.

5. LIBRARY AND INFORMATION MANAGEMENT SPECIALTY WORLDWIDE

For the purposes of this research we have studied the higher education institutions in the world, which provide training in "Library and Information Management" (LIM). In this study, we aim to gain information in which educational and qualification degree is offered training in the specialty LIM and what are the main components in the content of the curriculum. As a source of information have been used official websites of the respective universities. The results are presented in a synthesized form in Table2.

Table 2 . Worldwide Universities engaged in training in the specialty "Library and Information Management".

Country	Name of the University	Educational and Qualification Degree	Core Curriculum Elements
Australia	University of South Australia	Master (MA)	- Information Management - Accessing and Organizing Resources - Information Technologies
Canada	The University of British Columbia (School of Library, Archival and Information Studies)	Master	- Organization and Knowledge Management - Information Systems - Digital Libraries
China	Peking University	Bachelor (BA) Master	- Information Storage and Retrieval - Computer Networks - Information Economics - Management Information System - Library Management - Information Policy and Law
China	The University of Hong Kong	Bachelor Master	- Information Science - Science in the Field of Management
Denmark	Royal School of Library and Information Science	Master	- Theory of Production of Information and Knowledge - Analysis and Management of Information Resources - Information Policy and Strategy of the Organization - Planning, Management and Evaluation of Information Systems
Germany	Hochschule der Medien in Stuttgart	Bachelor Master	- Management in Human Resources, Organization, Marketing - Information and Media Market - Library Management System Database
India	Gujarat University	Master	- Management of Library and Information Centers - Leadership and Change Management - Information and Communications Technologies and Research
Malaysia	University Technology Mara	Bachelor	- Information and Library Management - Information, Information Technology and Libraries
Russia	Moscow State University of Culture and Arts	Bachelor Master	- Library and Information Management - Library Statistics, Legislation - Economics and Marketing of Library and Information Services - Information Technology Management - PR and Advertising in the Library
United Kingdom	University of Rhode Island (School of Library and Information Studies)	Master	- Management of Library and Information Services - Information Science and Technology - Types of Libraries and Library Processes and Services
	Robert Gordon University	Master	- Information Studies - Managing Information Services - Database Construction and Use
USA	University of California Berkley	Master	- Knowledge Organization - Information and Communications Technologies - Leadership - Computer Sciences - Psychology and Sociology, Economics, Business, Law, - Library/Information Studies

Table 2 shows that the universities conduct training in specialty 'Library and Information Management' mainly in master educational and qualification degree. The exceptions are Hochschule der Medien in Stuttgart, Germany; Peking University and The University of Hong Kong, China; Moscow State University of Culture and Arts (Department Information and Library Management), Russia; University Technology Mara (Faculty of Information Management), Malaysia.

Practical realization of bachelors graduated in 'Library and Information Management' in Bulgaria, is particularly suitable for libraries in small towns, whose staff is limited (and often reduced to a single librarian) and it is expected to be informed and competent in a wide range of issues. As answer of these needs of library and information sphere, SULSIT offers training in Bachelor level—specialty "Library and Information Management" and in Master level—"Library, information and Cultural Management".

Diplomas and certificates that graduates of the corresponding universities receive are different, account the type of the training already acquired specialty and professional experience and length of the training course.

They can be:

- Bachelour diploma (training lasts 3 - 4 years);
- Graduate sertificate (training is done after obtaining a bachelor's degree or professional experience and usually lasts 6 - 8 months);
- Postgraduate diploma (training is done after obtaining a bachelor's degree or professional experience and usually lasts for 12 months);
- Master (training is done after obtaining a bachelor's degree and usually lasts 16 - 24 months);
- Professional master (training is done after obtaining a bachelor's degree and professional experience and usually lasts 12 - 18 months);
- Graduate entry master (training is done after obtaining a bachelor's degree in Library and Information Management or related specialty and usually lasts about 24 months);
- PhD and other Doctoral Research Degrees (training is done after obtaining a master's degree or completion of a bachelor's degree with special honors and usually lasts 2 - 4 years).

On the development of educational content of specialty "Library and Information Management" in SULSIT

Table 3 . Partner universities of the Library Management Department in SULSIT, who carry out training in Library Management and/or Information Management.

Country	Higher School	Specialty and Level	Core Curriculum Elements
Austria	University of Vienna	Library and Information Studies **MA**	- Management (in Library, Information, Documentation) - System for Information and Documentation
Croatia	University of Zagreb	Library Science **BSc, MA, PhD**	- Bibliography - Computer Science - Library Management - Information Institutions Management Fundamentals - Libraries and Society
Czech Republic	Masaryk University Brno	Information and Library Studies **BA, MA, PhD**	- Library Processing and Services - Intellectual Property and Information Activity - Information Science - (Information Security, Information Systems, Information Education)
France	Institute of Technology Paris Descartes	DUT **Universal Technical Diploma**	- DUT Information Communication - Book Trade
Germany	Humboldt University in Berlin	Library and Information Sciences **BA, MA** Information Management **BA, MA**	- Digital libraries - Bibliometry, Infometriya, Scientometrics, - Information Policy, Ethics, Law - Information Retrieval and Exchange of Information - Communication and Management of Information and Knowledge
Germany	Hochschule der Medien in Stuttgart	Library and Information Management **BA, MA**	- Management in Human Resources, Organization, Marketing - Information and Media Market - Library Management System Database
Hungary	University of Szeged	Library and Information Science **BA, MA**	- Library Organization, Library System - Management of Technological Resources - Document Management - Legal Institutions of Documentation.
Lithuania	Vilnius University, Institute of Library and Information studies	Information Management in Libraries **BA, MA, PhD**	- Information and Library Service - Information Science - Culture Information and Communication - Creative Communication - Business Information Management - Public Relations
Poland	Jagiellonian University in Krakow,	Management and Social Communication **BA, BSc, MA, PhD**	- Information and Library Sciences - Culture and Media Management - Electronic Information Processing - Management - Social Policy
Turkey	Hacettepe University in Ankara	Information Management **BA, MA, PhD**	- Information Management - Information Architecture - Records and Database Management - Information Literacy, Standards and Cooperation in Information Management.

important influence had our joint work in the frame of Erasmus Intensive Programmes with our partners—the leading European universities, engaged in similar training programs in the field of LIS education. Table 3 presented the basic information about our partner's programs.

Review of the main directions in which are trained future library and information managers can generally be limited to the following: Information and Communication Technology and Resources; Information and Library Management (in particular—Management of Information Services, Management Information Systems, Personnel Management);

Knowledge Organization. Academic subjects included in these directions are different for different universities.

6. FINDINGS FOR FUTURE CURRICULUM REVIEW

A detailed analysis of the information from this research leads to interesting conclusions that will be very useful for future curriculum review of the "Library and Information Management". We achieved some findings and solutions that we suggest to the attention of the LIS academic community worldwide as useful and relevant to the context of modern educational paradigm.

Special attention in this article we would like to devote to the fact that some academic subjects which are included in the curriculum of SULSIT and met student's interest do not exist or are very limited in the curricula of universities abroad. The development and integration of these disciplines in our educational plan is provoked by the interaction with prospective employers that define this knowledge as important and necessary for library and information professionals.

We draw attention to the following subjects:

- Intellectual Property;
- Standardization in Library Activities;
- Quality Management in Library and Information Activities;
- Library Psychology;
- Bibliotherapy.

6.1. The Training on Intellectual Property

In an knowledge based economy, an important place take experts who can interpret issues related to intellectual property, such as librarians and information specialists. It is they who are responsible for creating a policy of promoting understanding and resolving legal disputes and conflicts that are unique to this aspect of the Information Society [16] . One way to achieve this is through the educational impact of the curriculum on intellectual property. Intellectual property can be considered as an element of information literacy in university information environment, so to develop successfully students at the university, and in life, they must learn to use efficiently and effectively the wide variety of information and communication technologies for searching, finding, organizing, analyzing and evaluating the information they need. In addition, they need to understand the ethical use of information, including the violation of individual rights to intellectual property as plagiarism, use without permission of the author of works of literature, art, science, and also of patented inventions, industrial designs, indications (trademarks, geographical indications, domain names, companies). Finally, they should be able to systematize all this knowledge together to create an effective final product. This requires them to assemble the entire package of basic skills for research, technological skills, critical thinking and evaluation.

6.2. The Training on Standardization in the Library Activities

Knowledge of standardization and specific standards in all areas of human activity are not only useful but also necessary in the direct work of specialists and leaders in different organizational structures, including libraries. Learning of this discipline implies a successful career of students and saves them a lot of effort, time and resources to achieve and implement the requirements, rules, standards, approaches, methods and etc., established by the most highly qualified professionals at international, European and national level and involved in the development of these standards.

Among the most common library standards that future library managers in SULSIT learn about can be mentioned:

- BGS ISO 11690:2013 Information and documentation—Library performance indicators.
- BGS ISO 2789:2012 Information and documentation—International library statistics.
- BGS ISO 15511:2012 Information and documentation—International standard identifier for libraries and related organizations (ISIL).
- BGS ISO 5127:2009 Information and documentation—Vocabulary.
- BGS ISO 2709:2011 Information and documentation—Format for information exchange.
- BGS ISO 25577:2012 Information and documentation—MarcXchange.
- BGS ISO 3166-1:2006 Codes for the representation of names of countries and their subdivisions—Part 1: Country codes.
- BGS ISO 11799:2008 Information and documentation—Document storage requirements for archive and library materials.

Standardization in Library Activities discipline, taught in the sixth semester is fundamental for studying in the seventh semester Quality Management in Library and Information Activities.

6.3. The Training on Quality Management in Library and Information Activities

The educational content in this course is structured in two parts:

Fundamental part, which includes the following topics: the importance and relevance of quality management; historical development, total quality management, principles of quality management, principles for analysis, security, control, organization, systems for quality management standards, quality management.

Today more libraries develop and implement systems for quality management. The most commonly used worldwide are those based on a series of international standards ISO 9000 [17] . Therefore lead in the course lectures Quality management in library and information work is devoted to this particular system.Are also presented two other systems for quality management application in libraries—Balanced Scorecard (BSPE) and selforganizations under the criteria of the European Foundation for Quality Management (EFQM).

Specialized part, consistent with the professional field in which to be realized students in "Library and Information Management." This part of the curriculum content provides knowledge, documentation and implementation of systems for quality management in libraries and information centres, in view of their professional activity. The seminars helped to give young professionals skills to formulate the mission of the library, and the ensuing quality policy and objectives to achieve quality. Special emphasis on the brainstorming technique of K. Ishikawaas an opportunity to identify the causes of a problem and their elimination [18] . Students are introduced to the documentation requirements of the quality management system, and sample content of the basic documents of the Quality Management System (QMS), which adapts to the specific type of library.

6.4. The Training on the Library Psychology

The problem of improving the quality of library services at the level of librarians' professionalism today is extremely topical. These factors increase their need for psychological and pedagogical knowledge.

The course on Library Psychology aims to equip students with true understanding and knowledge of the nature, functions and principles of Psychology, as part of the activities in the library environment to provoke new ways of working and communicating with users of library services. The course content is divided into two main modules. In the first module "Library Psychology" is considered library psychology as a scientific and academic discipline, traced the historical development of library psychology, the role of books in modern society. In the second module "Psychology of Communication in the Library" focuses on issues of personal and professional qualities of the library specialist; the psychology of communication between Librarian and Reader; the peculiarities of psychological climate in the library team and theformation steps of an effective team in the library; the nature and style of the library manager. Library Psychology has significant potential, not only theoretically, but practically oriented knowledge useful in the work of librarians from different types of libraries.

6.5. The Training on the Bibliotherapy

The course provides theoretical and practical knowledge to the students about the nature and application of Bibliotherapy in libraries. Students learn the historical and scientific development of Bibliotherapy. The Bibliotherapy brings together at least three fields of knowledge—a Science Studying the Book (Book Science, Literature, Library), the Science of the Human Soul, Targeted Bibliotherapy (Medicine, Psychiatry, Psychotherapy, Rehabilitation) and the Science of reading, (Ensuring Efficiency of Bibliotherapy, Psychology of Reading, Reading Instruction), which makes it extremely useful in the training of students. The lectures aim to examine possibilities which reveal Bibliotherapy by presenting its various forms, methods and functions. The students learn about new and alternative forms of employment applicable in the library environment.

The course includes the following three aspects: first module "Origin and Nature of Bibliotherapy" which focuses on the scientific development of bibliotherapy, definition of bibliotherapy, determining bibliotherapy, the types and purposes, her functions and tasks. In the second module, "Applications of Bibliotherapy" are tracked innovative techniques used by library professionals, bibliotherapy as a form of working with children with special needs. The third module, "Fairy Tales Therapy" is a brand new in Bulgaria and for the first time available as part of an academic course. In it, students get acquainted with the possibilities of therapy tales methodological model of working with stories, kind words and ways of working with them.

Bibliotherapy is dynamic and flexible science for support, encouragement, challenge, identifying and addressing a number of significant problems in modern society. It is a mechanism for painless reach the soul and the possibility of achieving lasting results and as such may be extremely useful in the training of future specialists.

7. DISCUSSION

In the present article, we pay attention to the above mentioned disciplines, as the analysis of the curricula of many universities around the world has shown that these disciplines are still rarely studied. In Table 2 and Table 3, Universities which studied one or more than one of the subjects on which we focus are present. Relatively new disciplines are presented by us, consistent with the requirements of the modern information society, and the role and functions of libraries in it. We believe that along with the subjects proved important for the preparation of future LIS professionals as library management, library legislation, library marketing, etc. would be appropriate curricula to attend and complementary knowledge of library professionals, consistent with the current state of society and libraries. The information presented can be used by other universities in Europe and beyond for comparison and analysis.

8. CONCLUSION

Through active partnerships with foreign universities, innovative teaching methods appropriate to the training of future specialists disciplines, and the continuous enrichment capacity and experience of the faculty, trained students in specialty "Library and Information Management"—we aim to be one of the leading educational institutions and adequate to the modern needs. We hope that shared insights and achievements will provoke an open professional dialogue for the enrichment of library and information academic programs. This collaboration could be going on together with the LIS community dialogue about IFLA Trends and the future of libraries and librarian profession in general [19] .

Acknowledgements This publication has been realized under the project "Development of an information environment to motivate and stimulate young researchers in SULSIT" by Contract: BG051PO001-3.3.06-0055. The project is funded by the Operational Programme "Human Resources Development" SchemeGrantsBG051PO001-3.3.06 "Support for the development of PhD students, post graduate students and young scientists" funded by the European Social Fund of the European Union.

REFERENCES

1. Kajberg, L. (2008) The European LIS Curriculum Project: Findings and Further Perspectives. Zeitschrift für Biblio-thekswesen und Bibliographie, 55, 184-189.
2. Juznic, P. and Badovinac, B. (2005) Toward Library and Information Science Education in the European Union: A Comparative Analysis of Library and Information Science Programmes of Study for New Members and Other Applicant Countries to the European Union. New Library World, 106, 173-186.
3. Georgy, U. (2009) Internationalization of LIS Education within the Bologna Process—Mobility and Flexibility. IFLA, Milano. http://www.uas7.org
4. Kawalec, A. (2014) Education, Competencies, Skills in the Field of Information and Library Science in Europe. Library (R)evolution: Promoting Sustainable Information Practices: Abstracts of the 22nd International BOBCATSSS symposium, 29-31 January 2014, Barcelona, 16. http://bobcatsss2014.hb.se/wp-content/uploads/2014/01/Bobcatsss-abstract-book.pdf
5. LIS Education in Europe. http://www.iva.dk/LIS-EU/project.asp/
6. Kajberg, L. and Lorring, L. (2005) European Curriculum Reflections on Library and Information Science Education. The Royal School of Library and Information Science, Copenhagen, 241 p. http://www.library.utt.ro/LIS_Bologna.pdf.
7. Pannier, G., Wilhelm, H. and Todorova, T. (2011) IPBib: Das Grimm-Zentrum-(k)ein Bibliotheksmarchen. Mobility and Innovation in the European Context. Proceedings of Evaluation Conference on ERASMUS Intensive Programmes, Federal Ministry of Education and Research, Bonn, 2011, 19-21.
8. Project Website (2013) Erasmus Intensive Programme "Library, Information and Cultural Heritage Management— Academic Summer School". http://libcmass.unibit.bg/
9. Todorova, T. (2012) Library, Information and Cultural Heritage Management: Textbook. Za bukvite-O Pismeneh, Sofia, 246.

10. Todorova, T., et al. (2013) The Changing Role of the Manager in the Digital Era: Findings from Erasmus IP LibCMASS 2012 Project in from Collections to Connections: Turning Libraries "Inside-Out". Proceedings of the 21th International BOBCATSSS Conference on Information Science in Ankara, Ankara, 23-25 January 2013, 202-204. http://bobcatsss2013.bobcatsss.net/proceedings.pdf
11. (2013) UNESCO Chair ICT in Library Studies, Education and Cultural Heritageat State University of Library Studies and Information Technologies. http://unesco.unibit.bg/
12. Yankova, I. (2009) Library Management and New Challenges in Digital Era in Libraries and Their Clients: Free or fee Services Supporting Social Communication in Digital Era. Proceedings of the ePublications of the 15th Jubilee International Conference of JU LIS Institute in Kraków, Poland, 76-80. http://skryba.inib.uj.edu.pl/wydawnictwa/e06/yankova.pdf
13. State University of Library Studies and Information Technologies. http://www.unibit.bg/learning-activity/bachelor/bachelor-plans.
14. Parizhkova, L. (2013) The Book Our More Sensual Present: Textbook. Za bukvite-O Pismeneh, Sofia, 248.
15. Trencheva, T. and Eftimova S. (2013) Intellectual Property at the Universities—Creativity: The Next Generation. Textbook. Zabukvite-O Pismeneh, Sofia, 206.
16. Joint, N. (2006) Teaching Intellectual Property Rights as Part of the Information Syllabus. Library Review, 55, 330-336.
17. Balague, N. and Saarti, J. (2011) Managing Your Library and Its Quality: The ISO 9001 Way. Chandos, Cambridge.
18. Ishikawa, K. (1990) Introduction to Quality Control, 3A Corporation, Tokyo.
19. (2013) IFLA Trend Report. http://trends.ifla.org/

12

CHAPTER

INTEGRATED INFORMATION ACCESS TECHNOLOGY FOR DIGITAL LIBRARIES: ACCESS ACROSS LANGUAGES, PERIODS, AND CULTURES

Biligsaikhan Batjargal,[1] Garmaabazar Khaltarkhuu,[2] Fuminori Kimura[1] and Akira Maeda[1]

[1] *Ritsumeikan University*

[2] *Mongolia-Japan Center for Human Resources Development*

INTRODUCTION

Physical libraries store materials written in various languages, at various periods in history, and dealing with various cultures. As a result, large digital library projects such as Europeana, World Digital Library, HathiTrust, and Google Book Search have collections spanning different languages, periods, and cultures. This diversity complicates information access, in part because the grammars, vocabularies, and scripts of languages usually change significantly over time.

This chapter presents our approach to providing cross-language access that accounts for this evolution of languages over periods ranging from ancient to modern and even considers cultural differences. It also presents our method for providing integrated access to multiple digital libraries, archives, and museums by automatically mapping between different metadata schemas. In section 2, we present the traditional Mongolian script digital library. Our proposed method for Cross-period information retrieval from ancient Japanese historical Materials is discussed in section 3. Later, in section 4, we introduce the federated searching system for humanities databases using automatic metadata mapping.

TRADITIONAL MONGOLIAN SCRIPT DIGITAL LIBRARY

In recent years the importance of digital cultural heritage preservation has been increasing in the Asia-Pacific region as well as worldwide. This section provides a summary of the recent achievements of the Traditional Mongolian Script Digital Library (TMSDL) (Khaltarkhuu et al, 2007; Khaltarkhuu et al, 2008), which aims to preserve over 800 years of historical records written in traditional Mongolian for future use and to make them available for public viewing. There are over 50,000 registered manuscripts and historical records written in traditional Mongolian script stored in the National Library of Mongolia. About 21,100 of them are handwritten documents and over 9400 of those are related to the history of Mongolia (Tungalag, 2005). Despite the importance of keeping old historical materials in good conditions, the Mongolian environment for material storage is not suitable for keeping historical records for a long time (Tungalag, 2005). An efficient and effective way to preserve and protect materials of historical importance while making them publicly available is to digitize them and create a digital library.

Mongolian is spoken by most of the Mongolian population as well as by Inner Mongolians and other groups of people who live in several provinces of China and the Russian Federation. It is one of the many languages of the Mongol-Altaic family.

Mongolians have used numerous writing systems, and traditional Mongolian script is the longest surviving script. Although the Mongolian language has also been written in Chinese characters, Phags-pa script, Soyombo script, Horizontal square script, Latin, and Cyrillic script (Shagdarsuren, 2001), by the end of the 20th century the traditional Mongolian script had made an official, government-decreed return. At the result, modern Mongolian language has two distinct writing systems: Cyrillic and traditional Mongolian.

The sounds of words changed as the Mongolian language evolved, but the spelling remained unchanged. Thus was created a difference between written Mongolian and spoken Mongolian. However, in 1946 in Mongolia the Cyrillic script was adopted with two additional characters. At that time the spelling of modern Mongolian in the Cyrillic alphabet was based on the pronunciation of the dialect spoken by the Khalkha, a subgroup of the Mongols. This was a radical change and alienated the Mongolian people from their culture and historical archives written in traditional Mongolian script. Traditional Mongolian script preserves a more ancient language and reflects the Mongolian language spoken in the ancient period, while modern Mongolian reflects pronunciation differences in modern dialects. Traditional Mongolian is a distinct dialect with grammar different from that of modern Mongolian. The traditional Mongolian script is written vertically, from top to bottom, in columns advancing from left to right. This script is the writing system for the Mongolian language and has four derivative scripts: Todo, Manchu, Vaghintara, and Sibe (Xibe). The Todo script was used by the Oirats and Kalmyks, and the Manchu script was a writing system in the Qing dynasty. The Sibe script is used in Xinjiang, in the northwest of China. The Vaghintara script was used by the Buryats. Like Arabic, traditional Mongolian is a contextual script where letters are cursively joined and have initial, medial, and final presentation forms for the same letter. In most cases the letters join together along a vertical stem, but in the case of certain consonants that lack a trailing vertical stem they may form a single ligature with a following vowel. In addition to these cursive and positional forms, many letters also have variant forms used in accordance with spelling and grammatical rules.

Using modern Mongolian to retrieve information from traditional Mongolian documents is not a simple task because the Mongolian language has changed substantially over time. The traditional Mongolian script digital library (TMSDL) (Khaltarkhuu et al, 2007; Khaltarkhuu et al, 2008), which is based on Greenstone Digital Library Software (GSDL) and accepts modern Mongolian query input, will help the user access materials written in traditional Mongolian.

Ancient-to-modern information retrieval

Thanks to advances in innovative information technologies and to the popularity of the Internet, many ancient historical documents are being digitized and made publicly available. We therefore want to offer an "ancient-to-modern information retrieval" method (Batjargal et al., 2010a; Batjargal et al., 2010b) that considers language differences over time. We aimed to develop a retrieval system with which a user can access cross-period and cross- script ancient document databases by using a query in a modern language.

There has been little research on information retrieval techniques for historical documents, and almost none of the breakthroughs in research on information retrieval and information access have aimed at retrieving information in the native language from ancient, cross- period and/or cross-script foreign language documents.

Few approaches that could be considered a cross-period information retrieval have been proposed (Ernst-Gerlach & Fuhr, 2007; Koolen et al., 2006; Gotscharek et al., 2009; Hauser et al.,2007; Pilz et al., 2008), and there has been little research on information retrieval techniques for historical documents. (Ernst-Gerlach & Fuhr, 2007) focused on modern and archaic German and developed a retrieval method that considers the spelling differences and variations over time. (Koolen et al., 2006) considered the spelling and pronunciation differences between ancient and modern Dutch, while (Gotscharek et al., 2009) and (Hauser et al., 2007) considered the spelling differences and variations between modern and archaic German. (Pilz et al., 2008) considered spelling variations of English and German historical texts. In general, the main challenge for historical European languages like Dutch, English, and German is the spelling variants.

We applied an "ancient-to-modern information retrieval" method to ancient Mongolian historical collections written in traditional Mongolian script. Some ancient historical documents in traditional Mongolian script have recently been digitized and made publicly available, and text-display support for traditional Mongolian script and the input locale is enabled in Windows Vista and Windows 7. The Uniscribe–Unicode Scripts Processor driver was updated to support OpenType advanced typographic functionality of complex text layouts, such as traditional Mongolian script.

The situation for an ancient Mongolian language is a bit different because the Mongols have changed their writing systems several times and more than once have made language reforms that eliminate a difference between written and spoken language (Shagdarsuren, 2001).

Proposed approach

To cope with cross-period and cross-script Mongolian documents, we propose a simple model that retrieves traditional Mongolian documents using modern Mongolian query. The structure of the TMSDL (Khaltarkhuu et al, 2007; Khaltarkhuu et al, 2008), with the proposed "ancient-to-modern information retrieval" approach (Batjargal et al., 2010a; Batjargal et al., 2010b) is shown in Fig. 1. We utilized the existing approach (Kimura et al., 2009) and improved the "retrieval technique with the modern Mongolian query on traditional Mongolian text" (Khaltarkhuu et al, 2006) by integrating a dictionary. A prototype of the TMSDL (Batjargal et al., 2010a; Batjargal et al., 2010b), which could be considered a cross-period information retrieval system, has been developed. The retrieval method of the TMSDL considers cross-period differences in the writing systems of the ancient and modern Mongolian languages. Adding a dictionary-based query translation approach to the translation module was a major improvement that takes into account age differences in the writing systems of the ancient and modern Mongolian languages. We utilized the developing online version of Tsevel's concise Mongolian dictionary (Tsevel, 1966) under the Creative Commons Attribution-Noncommercial-Share Alike 3.0 Unported license. Tsevel's dictionary was printed in 1966 and is one of two Mongolian dictionaries with definitions written in modern and traditional Mongolian available on the market. It includes over 30,000 words in Cyrillic and traditional Mongolian script.

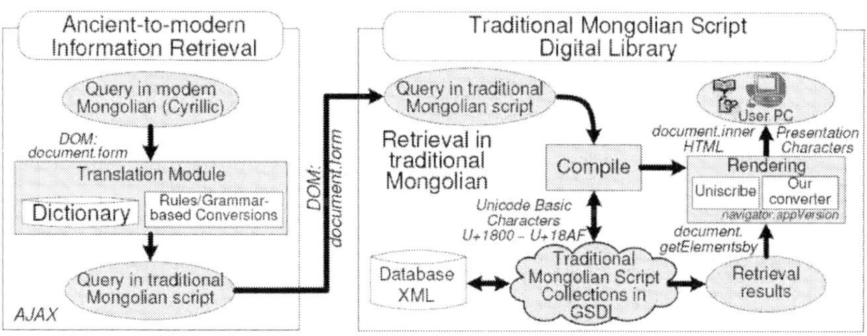

Figure 1. Ancient-to-modern information retrieval in the TMSDL.

To boost the quality of the translation, the "ancient-to-modern information retrieval" approach (Batjargal et al., 2010a; Batjargal et al., 2010b) matches query terms to words in Tsevel's dictionary. If no exact match is found, the "retrieval technique with the modern Mongolian query on traditional Mongolian text" (Khaltarkhuu et al, 2006), which is based on grammatical rules, is used. The proposed model allows the users to access documents written in an ancient language (traditional Mongolian) with a query input in a modern language (modern Mongolian – Cyrillic). As shown in Fig. 1, the query in modern Mongolian (Cyrillic) is translated into a query in traditional Mongolian script. The query in traditional Mongolian (Unicode characters in the range U+1800 – U+18AF) is then submitted as a retrieval query for traditional Mongolian script collections. Chronological books of ancient Mongolian kings, Genghis Khan, and the Mongol Empire (the largest contiguous empire in history) such as the Altan Tobci (year 1604, 164 pp) and the Story of Asragch (year 1677, 130 pp) etc, are available in the TMSDL with a modern Mongolian input interface. A database of such historical records with a modern language query input will help someone conducting research on the history of the High Middle Ages understand 13th–14th century history of Asia. The modern Mongolian (Cyrillic) input in the TMSDL is illustrated in Fig. 2.

Experimental evaluation

In an experiment we conducted in order to check the correctness of translations from the modern language to the ancient one, we retrieved traditional Mongolian documents when using modern Mongolian query input in Cyrillic. Because of the large number of unfamiliar ancient proper nouns, terms, and their variants in ancient historical documents, we faced the challenge of measuring recall and precision as well as the challenge of defining relevant documents. To check whether a queries in modern Mongolian (Cyrillic) were translated correctly, we selected queries the most frequently appearing words that are pronounced or written differently in modern and traditional Mongolian and compared their word counts in the search results with the corresponding word counts in "Qad-un űndűsűn quriyangγui altan tobči –Textological Study" (Choimaa & Shagdarsuren, 2002). This textological study contains a detailed analysis of traditional Mongolian word frequencies in the Altan Tobci.

We compared the word count in the search results for two cases: one using only grammatical-rule-based translation, and the other additionally using a dictionary. The version with dictionary integration translated and retrieved 86% of the input queries, whereas the grammatical-rule-based version retrieved only 61% of the input queries. Even with the dictionary, however, 64% of the input queries in modern Mongolian did not match with a word count that was less than or greater than the actual number (frequency) because of possible errors of translation, grammatical inflection, and text digitization, or limitations of the indexer and retrieval function. Comparisons of the retrieval results are illustrated in Fig. 3, and detailed retrieval results for sample query terms are shown in Table 1 along with modern and ancient forms, their meanings, and the word counts. A retrieval result with the query word highlighted is shown in Fig. 2.

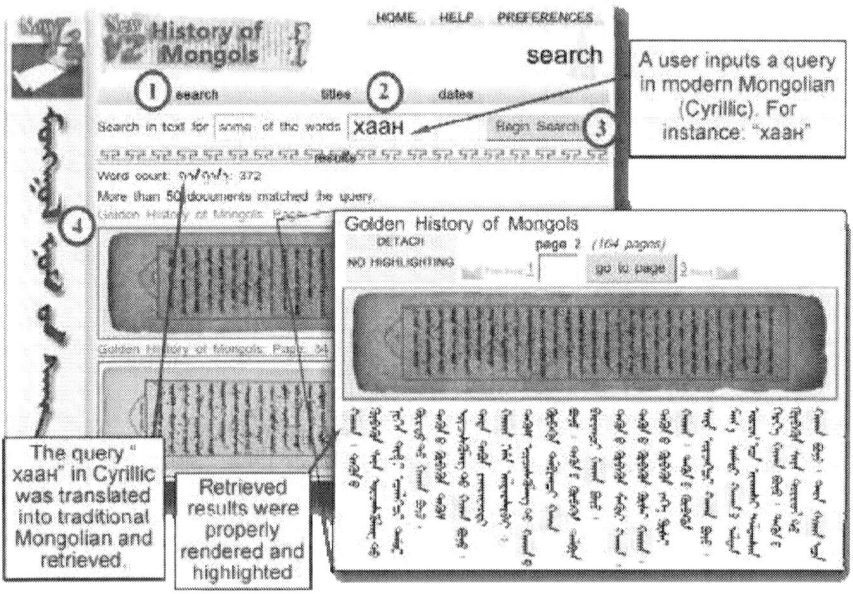

Figure 2. TMSDL Cyrillic input and retrieval results.

The TMSDL integrated with a dictionary translated and retrieved 86% of the input queries, but only 22% were retrieved without error.

Summary and future directions

In this section we introduced the TMSDL that utilizes cross-period and cross-script digital collections and that enables historical documents written in an ancient language to be accessed using a query in a modern language. The proposed system is suitable for full text searches on databases containing cross-period and cross-script documents. Such research would involve extensive research in an ancient language that users and humanities researchers may or may not understand. It could apply to humanities researchers who are conducting research on ancient culture and looking for relevant historical materials written in that ancient language. The proposed model will enable users and humanities researchers to search for such materials easily in a modern language. We still, however, need improvements dealing with such problems as a total failure to translate 14% of input queries. Improvements in translation and retrieval techniques also need to be considered.

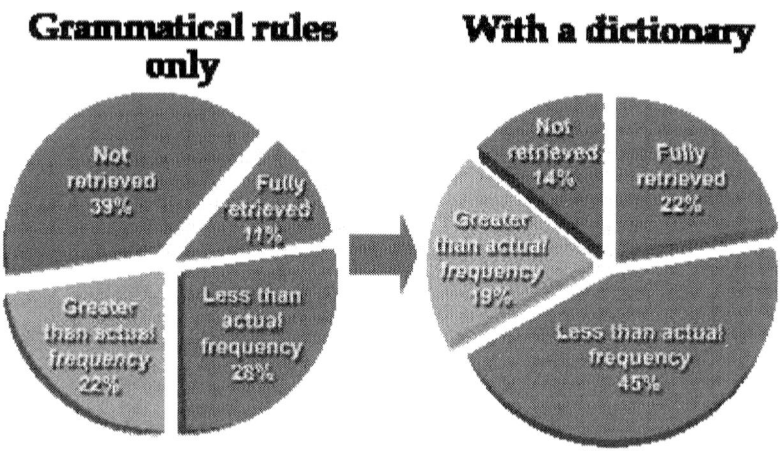

Figure 3. TMSDL retrieval results obtained using different translation methods.

Two interesting subjects for future research are the retrieval of information from two distinct ancient languages and using a single query input in a modern language for retrieval from multiple sources in multiple ancient languages.

The next section discusses the our another achievement – cross-period information retrieval method for ancient Japanese historical materials.

Table 1. Examples of retrieval performance obtained using different translation methods.

Word in Modern Mongolian (Cyrillic)	Pronunciation		Word in traditional Mongolian	English translation	Word count in retrieval results		Actual word count	Retrieval state
	Modern	Ancient			Grammatical rules only	With a dictionary		
хүн	hün	kümün	ᠬᠦᠮᠦᠨ	man, person, human	0	135	135	Fully retrieved
хаан	qaan	qaγan	ᠬᠠᠭᠠᠨ	king	0	372	372	
даян	dayan	dayan	ᠳᠠᠶᠠᠨ	all, whole	0	0	32	Not retrieved
сайн	sain	sayin	ᠰᠠᠶᠢᠨ	good, well, fine, nice, pretty	0	0	75	
гэр	ger	ger	ᠭᠡᠷ	home, residence	0	51	47	Greater than actual frequency
төр	tör	törü	ᠲᠥᠷᠥ	law, kingdom	0	50	47	
эзэн	ezen	ejen	ᠡᠵᠡᠨ	lord	0	144	146	Less than actual frequency
богд	bogd	boγda	ᠪᠣᠭᠳᠠ	holy, sacred, divine	0	39	40	

CROSS-PERIOD INFORMATION RETRIEVAL FROM ANCIENT JAPANESE HISTORICAL MATERIALS

Libraries, governments, and major internet providers have recently begun forming consortiums to preserve historical documents stored in libraries. This means that more and more old-text content will soon be accessible on the Internet. The huge amount of knowledge in old documents is obviously as important as that in the recently created digital documents typically available on the web because old documents contain the wisdom of our ancestors.

Retrieving important information from old documents is not always easy, however, because languages and cultures change substantially over time. To access documents written in ancient Japanese by using a query in modern Japanese, for example, we need a cross-period information retrieval system based on a cross-period (ancient-modern) Japanese dictionary.

Construction of ancient-modern dictionary

Ancient documents in text form are being digitized, and the prevalence of search engines has made the retrieval of information from digital documents a familiar procedure. Current search engines, however, may be not able to acquire proper retrieval results for ancient Japanese documents because there is no ancient-modern Japanese dictionary with sufficient entries.

One reason for this is that the Japanese writing system has no term separation. That is, neither current nor ancient Japanese writing uses space or punctuation to separate words. A morphological analyzer like ChaSen or MeCab, both of which need a modern term dictionary, is usually used to do term separation for modern Japanese, but there is no ancient-modern word dictionary with enough entries and there are no morphological analyzers for ancient Japanese. This makes it difficult to do term separation for ancient Japanese.

We propose a method for constructing an ancient-modern Japanese dictionary by using a parallel corpus of ancient writings and their translations in modern Japanese. The parallel corpus thus consists of pairs of documents in the same language but in ancient and modern versions of that language. From this corpus we try to acquire pairs of equivalent archaic and modern words by analyzing the frequencies of word occurrences in a sentence in ancient Japanese and its corresponding modern Japanese translation.

1.1.1 Related work

Two methods for extracting pairs of equivalent words from a bilingual corpus in modern languages (English and Japanese, for example) have already been proposed, one using a parallel corpus and the other using a non-parallel corpus. In the method using a parallel corpus, equivalence is based on statistical correlation determined using co-occurrence frequency, contingency tables, etc. (Kitamura & Matsumoto, 1996). In the method using a non-parallel corpus, equivalence is based on the context similarity of translation candidates (Tanaka, 2002). The method described here, however, identifies pairs of equivalent words not in two modern languages but in modern and archaic Japanese. As there are few modern language translations of ancient writings, it is difficult to collect a parallel corpus of ancient writings and their translations in modern language. Some famous ancient writings, though, have been translated into the modern forms of their languages. We therefore identify pairs of equivalent words in modern and archaic Japanese by using a parallel corpus comprising famous ancient Japanese writings and their translations in modern Japanese.

1.1.1 Proposed method of dictionary construction

Many well-known ancient writings have modern-language translations, and some of these translations are digitized and open to the public. In a parallel corpus comprising writings in an ancient and modern language, one can usually determine which modern-language sentence corresponds to which ancient-language sentence. A modern word equivalent to an archaic word in an ancient-language sentence is likely to appear in the modern-language translation of that sentence, and vice versa. Word pairs with high co-occurrence frequency in ancient and modern sentence pairs are thus likely to be translation equivalents.

In our method we detect similarities in the appearance tendencies of modern and archaic words in each sentence pair and then use these similarities to extract equivalent pairs of ancient and modern words (Fig. 4).

A. Word Extraction from Parallel Corpus

We use morphological analysis to extract words from the modern-language translations of the ancient writings, and because there is no morphological analyzer for ancient Japanese. We divide the archaic sentences into N-grams and treat those N-grams as archaic words.

An N-gram is a sequence of N characters from a given string. We first extract the first N characters from the target string and then shift one character and extract N characters from the target string. We repeat this shifting-and-extracting process until the Nth character in the N-gram is the last character of the target string. For example, the string "corpus" would be divided into the following four 3-grams: cor, orp, rpu, and pus.

One of the drawbacks of the N-gram approach is that there will be many overlaps. On the other hand, an advantage of the N-gram approach is that it can divide the strings even if the language of the string, like ancient Japanese, does not have explicit delimiters between words. This is why we divide the archaic sentences into N-grams and treat those N-grams as words.

B. Calculation of the Co-occurence of Modern and Archaic Words

In this process, we calculate co-occurrence frequencies of archaic terms and modern terms that are extracted in section 3.1.2.A. This process is conducted for archaic and modern term pairs to appear in the equivalent sentences. In other words, the term pairs appearing in the equivalent sentences are considered as the co-occurring terms.

In each sentence pair, the archaic and modern term pairs are created for every possible pairs of extracted modern terms and archaic N-grams. We count the occurrence frequency of each term pairs. This frequency is the co-occurrence frequency of archaic and modern term pairs.

C. Calculation of Similarity about Appearance of Tendency between Modern Term and Archaic Term

For parallel corpus composed two different languages documents such as Japanese and English, "mutual information" is proposed to use for the similarity between each two terms (Kitamura & Matsumoto, 1996). Our method also adopts "mutual information" in order to calculate similarities about appearance of tendency between modern term and archaic term. The archaic and modern term pairs that have higher value of their mutual information is considered that appearance of tendency between modern term and archaic term is similar. These term pairs have higher possibility that the modern term is relation in translation for the archaic term. We extract term pairs that have higher similarities than some threshold, and consider that these pairs have relation in translation.

The mutual information MI of a modern term t and an archaic N-gram g is given by the following formula.

$$\mathrm{MI}(t,g) = \log \frac{P(t,g)}{P(t)P(g)} \tag{1}$$

Probability $P(t, g)$ is the probability that the modern term t appears in the translation of the archaic sentence in which the archaic N-gram g appears, and it can be calculated from the co-occurrence frequency of archaic N-gram g and modern term t. Probabilities $P(t)$ means the probability in the case that the modern term t appears in modern sentence. Probabilities $P(g)$ means the probability in the case that the archaic N-gram g appears in archaic sentence. Probabilities $P(g)$ is able to be acquired from the term frequency of archaic N-gram g as mentioned in section 3.1.2.A.

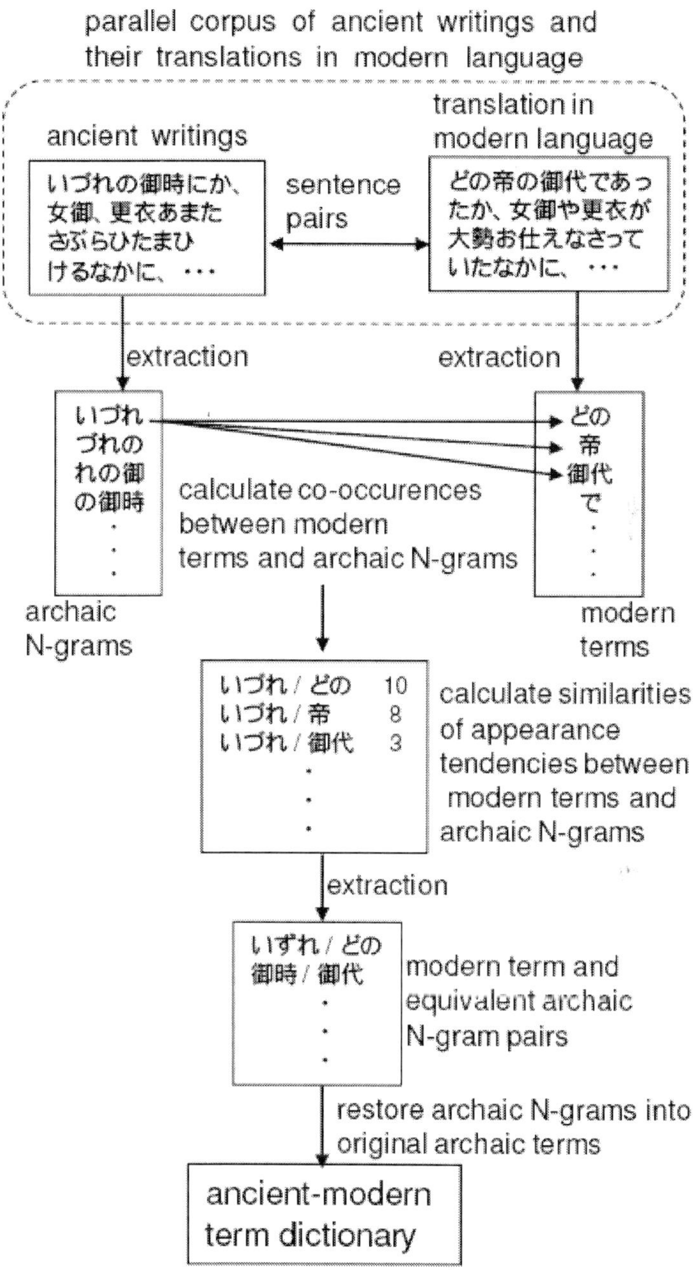

Figure 4. Flow of the construction of an ancient-modern dictionary.

The archaic and modern term pairs that have higher value of their mutual information are considered that appearance of tendency between modern term and archaic term is similar.

D. **Extraction of Translation Pairs of Modern and Archaic Words**

In section 3.1.2.C, we extract archaic and modern term pairs that have higher possibilities of relation in translation. However, as the archaic terms of extracted pairs are represented by N-gram, these archaic terms are not always complete archaic term. Some archaic N-gram may be part of archaic term. Another may be combined parts of some archaic terms. In these cases, we have to restore the archaic N-grams to original archaic terms. These archaic N- grams are restored to original archaic terms by comparing spellings, term frequency and co- occurrence frequency between another archaic N-gram. We consider that restored archaic and modern term pairs are related in translation. Finally, we collect these term pairs and construct ancient-modern term dictionary.

3.1.2 Future directions

We proposed a method for constructing an ancient-modern Japanese dictionary by using a parallel corpus of ancient writings and their translations in modern Japanese. If an ancient- modern Japanese dictionary with sufficient entries is constructed by the proposed method, we think that the techniques of natural language processing, for example morphological analysis, could be applied for ancient documents digitized in text form.

We need to improve the term extracting process in order to reduce the number of unnecessary word pairs, to improve the calculation of similarities of the appearance tendencies of modern and archaic words, and to construct a practical ancient-modern Japanese dictionary.

3.2 Cross-period information retrieval system

There has been a lot of research on cross-language information retrieval in the last decade. Various approaches—including query translation, document translation, and the use of an intermediate language—has been studied, and adequate retrieval effectiveness has been achieved for some pairs of languages (e.g., certain European languages).

There has, in contrast, been very little research on information retrieval methods for historical documents, and most of those methods are based on simple keyword matching. Some recently proposed approaches to accessing historical documents consider the evolution of languages and could be regarded as a kind of cross-age information retrieval (Gerlach & Fuhr, 2007; Khaltarkhuu & Maeda, 2006). Our goal is to establish a more effective and sophisticated retrieval method that considers not only language difference over time but also cultural differences between languages and ages.

The architecture or the cross-period information retrieval system we developed is shown in Fig. 5. This system lets old Japanese documents be retrieved using modern Japanese keywords, so old Japanese documents by users who do not know archaic Japanese.

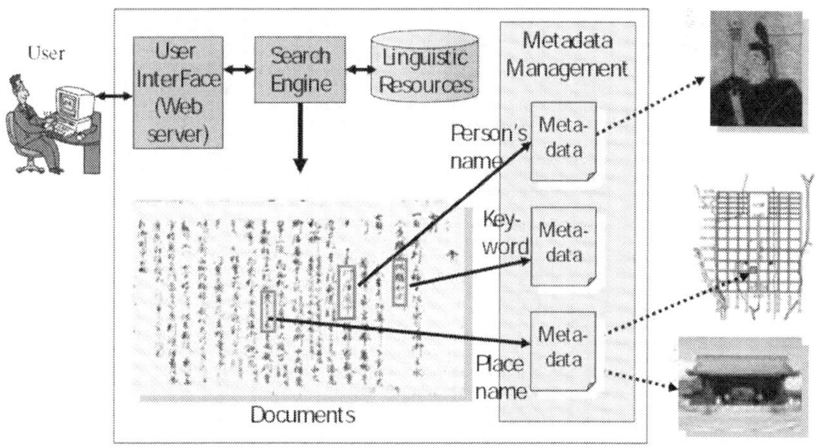

Figure 5. Architecture of proposed cross-period information retrieval system for ancient Japanese historical materials.

3.2.1 Proposed method for cross-period information retrieval

We use the dictionary-based query translation approach because it is the one most effective for cross-language information retrieval. For dictionary-based methods to be effective we need to use precise and comprehensive dictionaries for both the modern and ancient language. We try to find relations between the entries in those dictionaries and to translate the query terms in the modern language into equivalent terms in the ancient language. For this translation process we propose the following method (Fig. 6).

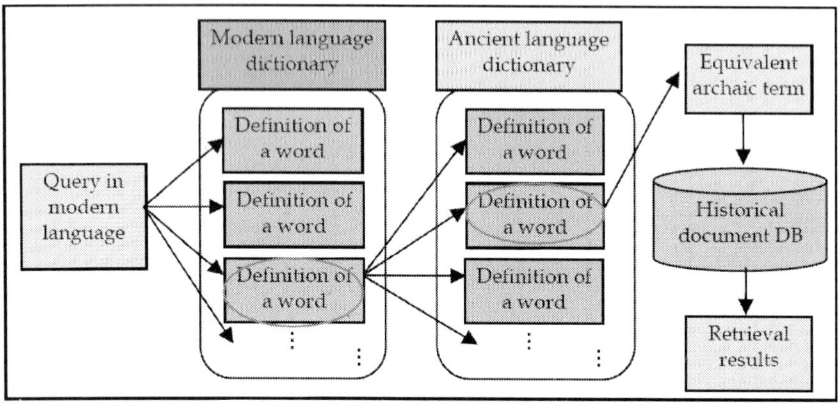

Figure 6. Overview of proposed method for cross-period information retrieval.

1. For each entry in the modern-language dictionary, we look for an equivalent entry in the ancient language dictionary by calculating the similarities between the definition of the modern word and all the definitions of the archaic words. We can do this using a standard text similarity measure based on the vector space model and the tf-idf term weighting scheme.

2. We then take the most similar definition in the ancient language dictionary and regard the dictionary entry (headword) containing that definition as an equivalent of the modern word.

3. If there is more than one equivalent entry, we find the one most nearly equivalent to the modern word by using a term association measure such as mutual information to disambiguate the candidate translations.

3.2.2 Implementation

We implemented a cross-period information retrieval system for the Japanese historical document called the *Hyohanki* diary. Written in late Heian era (12th century), it is a valuable resource for research on Japanese culture of that time. An example of its original copy is shown in Fig. 7. Part of the *Hyohanki* has deteriorated and is missing, but all of the existing pages (comprising 2,488 diary entries) have been digitized into text format.

As described in Section 3.2.1, we need dictionaries in order to translate modern language query words into archaic words. In the case of the *Hyohanki* diary we can use some existing elcctronic dictionaries available on CD-ROMs. For modern Japanese we use *Kojien*, one of the most famous and comprehensive Japanese language dictionaries. For ancient Japanese we use *Kokugo-Daijiten*, which covers not only modern words but also archaic words.

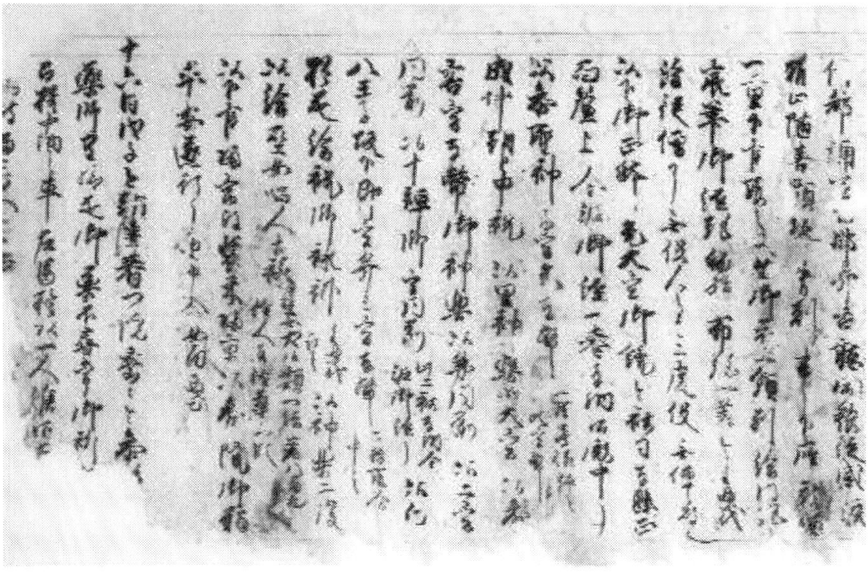

Figure 7. Part of the original copy of the historical Japanese document *Hyohanki*.

3.2.1 Experiment

We conducted a preliminary experiment to test the precision of cross-period retrieval by our proposed method. We used *Hyohanki* diary entries of as the ancient Japanese document

collection and prepared three modern Japanese queries: ચોl૪ (war), ◉ր (Buddhist service),

and ҕ⍰ (bare foot). Since the archaic equivalent of each query differs from the query itself,

no relevant documents can be retrieved if we use these modern term queries. Note that we consider one diary entry as one document.

Table 2 shows the original modern Japanese query, the ancient Japanese equivalents (translations) obtained by the proposed method, and the precision of retrieval using the

translations. For the queries ⊙ര (Buddhist service) and ҕ⍰ (bare foot), the proposed

method worked quite well: 99–100% precision (the ratio of relevant documents in retrieved documents). The query ઓlઠ (war), however, resulted in very poor precision (27%) because the proposed method returned two translation candidates for this query: ઓl and .. If we use only ઓl as the translated query we obtain 100% precision, but if we use only . we obtain only 3.6% precision. This is because the archaic term . has not only the meaning *war*

but also meanings like *general* (officer) and *army*. The query ⍰⍰ (death) also resulted in very poor precision (15%) because its translation ⍰ also means *deprivation* and *sunset*. These results suggest that we could improve the precision if we incorporate a suitable disambiguation method for the translated archaic terms. For that purpose, we could apply existing disambiguation methods used in Cross-language Information Retrieval.

Table 2. Precision of the retrieval results in cross-period retrieval.

Modern Japanese query	Translations	Relevant / Retrieved
戦争 (war)	軍, 戦	10 / 37 (27%)
法要 (Buddhist service)	仏事	109 / 110 (99%)
裸足 (bare foot)	跣, 裸足, 跣足	27 / 27 (100%)
死亡 (death)	没	2 / 13 (15%)

FEDERATED SEARCHING SYSTEM FOR HUMANITIES DATABASES USING AUTOMATIC METADATA MAPPING

This section provides a summary of our approach to constructing a federated searching system for Japanese humanities databases using automatic metadata mapping. The goals of our system are (1) to perform metadata mapping automatically for Japanese heterogeneous humanities databases and (2) to let users access multiple humanities digital libraries by using only one query input. This section also addresses the metadata-related challenges facing Japanese humanities databases. Metadata offers library and information science a solution to the problem of describing and managing the massive quantities of explosively increasing digital information (Zeng & Jian, 2008). Various types of resources and humanities digital libraries coexist with heterogeneous metadata schemas nowadays, and many different metadata schemas are standardized by international standards organizations. How to deal with the diverse forms of metadata and interoperate is becoming a complex issue for research. There have been efforts to make heterogeneous standards interoperable and utilize multiple metadata standards. According to (Chan & Zeng, 2006), several different approaches (element mapping, crosswalk, application profile, metadata registry, etc.) were developed. Reliable metadata interoperability has not been achieved yet because of the heterogeneity of metadata standards and because of the structural differences between standards.

On the other hand, the use of metadata schemas and standards for Japanese humanities digital libraries is a bit tricky. Many metadata schemas of Japanese humanities digital libraries have been accepted in terms of their semantics and content but were developed before the international metadata standards or were developed without considering the international metadata standards and specific encoding methods. Most of the metadata schemas of Japanese humanities digital libraries were not derived from existing international metadata standards, and there is no explicit metadata framework, crosswalk, or metadata registry. It is necessary to understand the semantics of Japanese humanities digital libraries—such as elements, syntax, and structure—in order to perform automatic metadata mapping and achieve metadata interoperability. This section therefore addresses the metadata-related challenges to constructing a federated searching system for Japanese humanities databases.

Metadata schemas for Japanese humanities digital libraries and their challenges Humanities digital libraries and their metadata schemas are very heterogeneous because the humanities cover a variety of disciplines, such as literature, law, history, philosophy, religion, visual and performing arts (including music), anthropology, cultural studies, and linguistics (including ancient and modern languages). Achieving metadata interoperability of humanities digital libraries is becoming more crucial in the current information environment, especially in the case of metadata schemas which were not derived from well- known international metadata standards.

One of the differences between western and Japanese databases that is relevant to people interested in constructing a federated searching system is the greater heterogeneity of the metadata schemas of Japanese humanities digital libraries. Many Japanese humanities databases developed metadata schemas based on their domain-specific semantics and content rather than adopt international metadata standards. Moreover, names or labels for metadata attributes/elements are written in Japanese, or labels in Japanese are used as the metadata elements. The co-existence of nonstandard and heterogeneous metadata schemas makes automatic metadata mapping for Japanese humanities databases a rather challenging task.

Another relevant difference is the Japanese writing system(s). Japanese is written in a mixture of three writing systems—one using ideographic symbols, or *kanji*, and the other two using the syllabary scripts *hiragana* and *katakana*—and it is written without explicit word boundaries. The absence of word delimiters makes word segmentation (i.e., tokenization) a critical problem in natural language processing for Japanese. Without knowing the boundaries of words in a sentence, any computer system will fail to perform tasks such as automatic metadata mapping. A single kanji can have many pronunciations and be used differently in words comprising two or more kanji. The situation will be much more difficult when collections contain ancient documents because a modern kanji is not always the same as its archaic equivalent. An archaic word written with a single kanji might be equivalent to a modern word written with more than a single modern kanji, or vice versa. Using a modern language query to find information in Japanese documents that are written in modern and archaic Japanese words is a rather challenging task.

Federated searching system for Japanese humanities databases

The conceptual architecture of our proposed federated searching system is shown in Fig. 8. As illustrated there, if a user wants to find a humanities resource with the query word in the title, our system retrieves resources having the query word in the title or any metadata field that is similar to a title or could be treated as a title and retrieves these resources from heterogeneous humanities digital libraries even if those libraries do not provide metadata interoperability or crosswalk and do not support Z39.50 protocol, Search/Retrieve Web service (SRW)/Search/Retrieve via URL (SRU), etc.

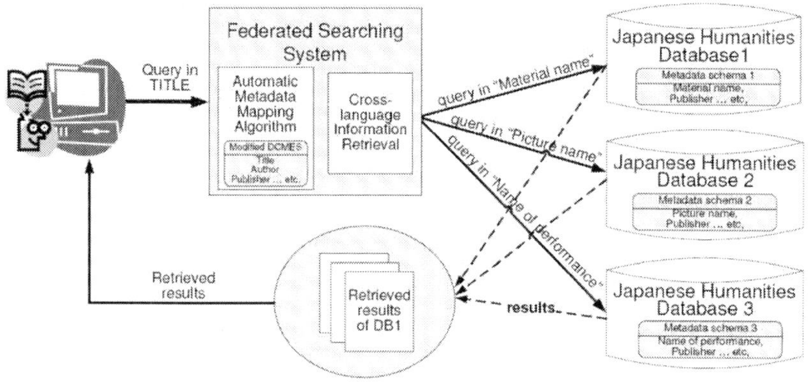

Figure 8. Conceptual architecture of the proposed federated searching system.

We are developing a prototype federated searching system of Japanese humanities databases—including the image database of Japanese traditional fine art Ukiyo-e, donated Japanese books database, and old Japanese books database—that are freely accessible in Japanese at the Art Research Center of Ritsumeikan University. We utilized the automatic metadata mapping method of Kimura et al. (2009). This prototype system also has a facility for cross-language searching between English and Japanese, which enables English- speaking users to search Japanese databases available only in Japanese.

4.1 Automatic metadata mapping

In our system the metadata attribute names of heterogeneous Japanese humanities collections in Japanese, the metadata schemas of which are unknown or do not conform to the international standards, are automatically mapped to our modified variant set (hereafter, modified DCMES) of the Dublin Core metadata element set (DCMES) (Dublin Core Metadata Initiative, 2008). Because CREATOR and CONTRIBUTOR are hard to distinguish in Japanese humanities collections, in the modified

DCMES they are unified into the new element AUTHOR. When Japanese humanities metadata schemas are successfully mapped to the modified DCMES, our proposed system enables cross-domain metadata harvesting and federated searches as well as the exchange of metadata.

Our automatic metadata mapping method (Fig. 9) consists of two preprocessing phases and four mapping phases.

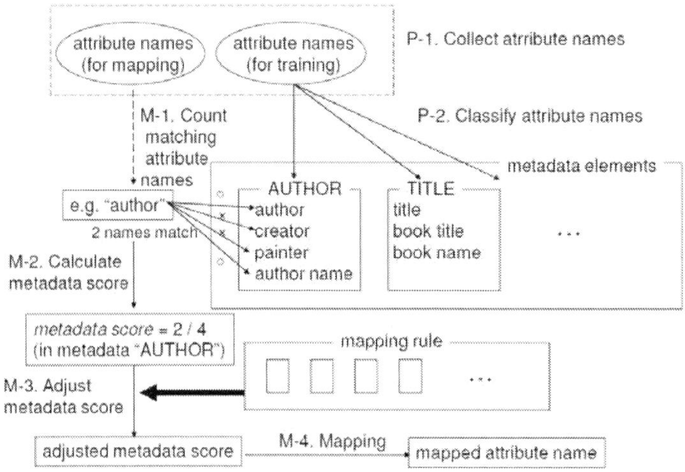

Figure 9. Flow of automatic metadata mapping.

The preprocessing consists of the following steps:

P-1 Collect attribute names from humanities databases for training and mapping.

P-2 Classify attribute names for training into appropriate metadata elements manually.

The automatic mapping phase consists of the following steps:

M-1 Count the number of partial string matches between the attribute name for mapping and each metadata element.

M-2 Calculate the *metadata score* of each metadata element by dividing the number of partial string matches by the number of attribute names in the metadata element.

M-3 Adjust the metadata score for each metadata element, if the target attribute name matches one or more *mapping rules*, which consist of some kanji characters (or partial words) that are commonly used and known to be relevant to one or more particular metadata elements. (e.g., increase the metadata score for "TEMPORAL" if the attribute name includes "year").

M-4 Map the target attribute name to the metadata element that has the highest metadata score.

If the attribute name is given the metadata score value 0 for all metadata sets, the attribute name is classified into "OTHER" metadata.

Table 3. Example of results of the automatic metadata mapping.

Modified DCMES elements mapping	Metadata elements of Japanese humanities digital libraries	Meanings of kanji used in metadata elements of Japanese humanities digital libraries	Number of elements
TITLE	画題等，画題2，役名，外題，外題よみ，所作題，所作題よみ，細目題，細目題よみ，主外題，主外題よみ，系統分類題，演目(統合)，演目よみ(統合)，画題統合，資料名，資料名よみ，解題	Print title, Picture name, Character names, Official title, Played title, Title of play, Reading of played title, Performed title, Reading of performed title, Detailed title, Reading of detailed title, Main performed title, Reading of the main performed title, Classification title, Name of performance, Reading of the performance, Title of the integrated picture , Material name, Reading of material name, Synopsis	18
PUBLISHER	版元文字，異版，版印1Nb，版元1Nb，版元1，版印2Nb，版元2Nb，版元2，版元備考，地域版，版元統合	Character publisher, Different edition, Edition stamp #1, Publisher #1, Publisher 1, Edition stamp #2, Publisher #2, Publisher 2, Publisher remarks, Domestic publisher, Joint publisher	11
DATE	西暦，和暦，年月日備考，月日--計算，西暦版，和暦版，月日版，年月日備考版，閏，月，日	Gregorian calendar, Japanese calendar, Edited date, Date calculation, Gregorian calendar edition, Japanese calendar edition, Edition date, Remarked date, intercalary, Month, Day	11
AUTHOR	絵師，編著者等，原所蔵者，彫師等，担当者	Artist, Volume author etc., Original owner, Engravers, etc., Person in charge	5
COVERAGE	地域，位置，続方向，劇場，場立，場名	Performed Place, Location, Spatial, Theater, Place, Place name	6

Inspecting the data listed in Table 3, one sees that 18 metadata elements (attribute names) of the Ukiyo-e image database, donated books database, and old books database were mapped to the TITLE element in the modified DCMES. Similarly, 11 elements were mapped to DATE, 11 to

PUBLISHER, 5 to AUTHOR, and 6 to COVERAGE. These eighteen attribute names were written in various kanji characters that have different meanings, such as "Print title," "Picture name," "Character names," "Official title," "Played title," "Title of play," "Reading of played title," and "Performed title The metadata attribute names used in Japanese humanities digital libraries consist of several words that have combinations of single or several kanji characters, and the meaning of the words depend on the combinations. Our algorithm performs automatic mapping by calculating the overall metadata scores for each metadata element, which are calculated for the words or kanji characters by using training data set and mapping rules. For instance, if the name of a

metadata element has the character [illegible] (name), increase the metadata score for TITLE by 1,

for PUBLISHER" by 0.5, and for AUTHOR by 1.

Our study of 334 metadata elements of 50 Japanese humanities digital libraries showed that 65 different elements have a potential to be regarded as TITLE, 46 as AUTHOR, 25 as SUBJECT, 77 as DESCRIPTION, 22 as PUBLISHER, 5 as TYPE, 20 as IDENTIFIER, 5 as

SOURCE, 44 as COVERAGE, and 7 as RIGHTS. This shows how heterogeneous metadata schemas of Japanese humanities digital libraries are and that is vital to perform metadata mapping automatically.

Modified DCMES elements	Average precision (%)
TITLE	89.9
SUBJECT	100.0
AUTHOR	91.8
PUBLISHER	85.7
IDENTIFIER	100.0

Table 4. Mapping precision of the automatic metadata mapping method.

Metadata Sets	Conditions	Average precision (%)
Standard Dublin Core Metadata Element Set	Without mapping rules	73.8
Standard Dublin Core Metadata Element Set	With mapping rules	79.0
Modified Dublin Core Metadata Element Set	With mapping rules	94.9

Table 5. Comparison of metadata mapping precision.

According to the judgement of a native Japanese speaker experienced in Japanese humanities digital databases who checked the results obtained when our automatic metadata mapping method mapped 334 attribute names of Japanese humanities collections to metadata elements of the modified DCMES, the average mapping precisions ranged from 85.7% to 100% (Table 4).

The average precisions we obtained using standard DCMES without the mapping rules, using standard DCMES with the mapping rules, and using modified DCMES with the mapping rules are listed in Table 5, where one sees that the mapping precision obtained using modified DCMES with the mapping rules is 21.1 percentage points higher than that obtained using standard DCMES without the mapping rules, and this shows that mapping rules improve the metadata mapping considerably. The average precision obtained using modified DCMES with the mapping rules was 15.9 percentage points higher than that obtained using the standard DCMES with mapping rules, and this shows that the modified DCMES also improves the metadata mapping considerably.

Retrieval in a federated searching system using automatic metadata mapping

To examine the performance of our federated searching system using automatic metadata mapping, we conducted an experiment by inputting a single query to three humanities collections (Ukiyo-e image database, donated Japanese books database, and old Japanese books database). Retrieval results obtained from three collections for the sample query (elegance) in the TITLE metadata fields are shown in Fig. 10. Retrieval with other sample queries was also successful.

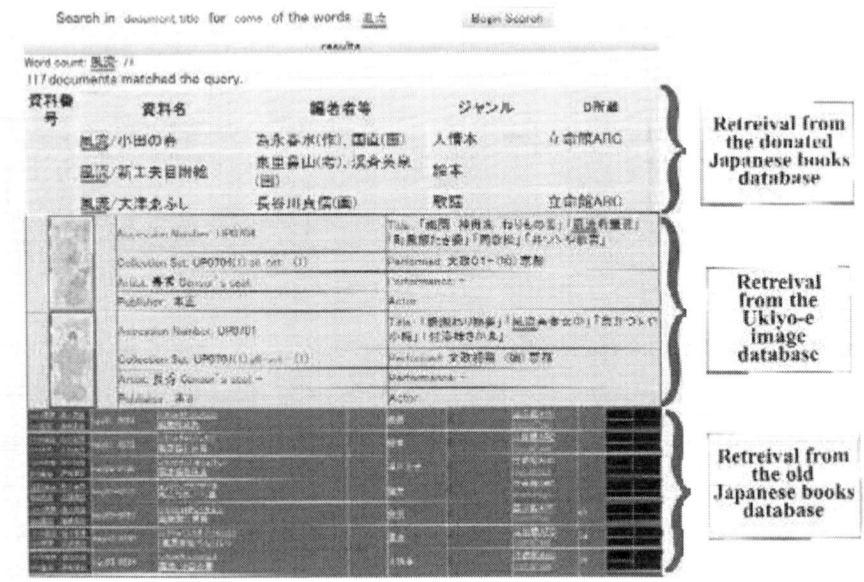

Figure 10. Retrieval results obtained from three from Japanese humanities digital libraries when using automatic metadata mapping.

Retrieval in a federated searching system using English queries

Our federated searching system also retrieves resources from Japanese collections when an English query is used. This feature is very useful for users who do not understand Japanese, and it allows searching and browsing Japanese digital libraries in English through a single interface and a single query (Batjargal et al., 2010c). We applied this feature to the Ukiyo-e image database of the Art Research Center of Ritsumeikan University, which is freely accessible in Japanese.

Ukiyo-e, Japanese traditional woodblock printing is known world-wide as one of the fine arts of the Edo period (1603–1868). The texts of Ukiyo-e databases contain archaic Japanese words which reflect the Japanese language of the Edo period. Besides providing information about Ukiyo-e prints, the Ukiyo-e database of the Art Research Center of Ritsumeikan

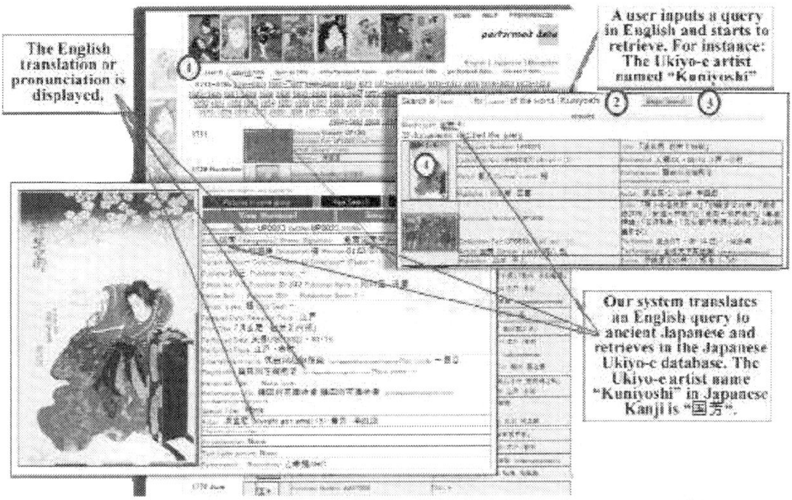

Figure 11. Using an English query to search Japanese Ukiyo-e databases.

University contains information about the content of the prints. For instance, if the subject of an Ukiyo-e print is Kabuki, the highly stylized classical Japanese dance-drama, the database contains some additional information. Sometimes explanations of cultural and social meaning for the print are also included.

67 metadata elements of the Ukiyo-e database are mapped to the modified DCMES using our automatic metadata mapping method. As shown in Fig. 11, the Ukiyo-e artist name

Kuniyoshi as an input query was translated as □ & and retrieved from the Japanese Ukiyo-e

image database. The translated terms, names, explanations, etc. were displayed in English pages. Multiterm queries were treated as words: the artist's full name *Utagawa Kuniyoshi*,

was treated as ч⍰ (Utagawa) and □& (Kuniyoshi) but not as ч⍰□&. As illustrated in

Fig. 11, users will be able to enter a query in English (2) after clicking the Search button (1). The query *Kuniyoshi* is translated as □& when the Begin Search button is clicked (3), and

the translated query is retrieved from the Japanese Ukiyo-e image database. Lastly, the user will be able to access the webpage (4) that displays detailed information of a certain Ukiyo-e print, where the metadata in Japanese are translated and displayed in English.

SUMMARY

In this chapter we presented some of our work related to integrated information access technology for digital libraries. We developed technologies providing information access across different languages, periods, and cultures. These technologies will be particularly important for large digital library collections that include contents written in different languages and spanning a wide range of periods and diverse cultures. The systems presented in this chapter were developed primarily for humanities researchers but might also be useful to ordinary users because much of the knowledge and wisdom in old documents is not available in modern-language documents.

ACKNOWLEDGEMENTS

This work was supported in part by the Grant-in-Aid for the Global COE Program "Digital Humanities Center for Japanese Arts and Cultures (DH-JAC)" from the Ministry of Education, Culture, Sports, Science and Technology (MEXT) of Japan, MEXT Grant-in-Aid for Strategic Formation of Research Infrastructure for Private University "Sharing of Research Resources by Digitization and Utilization of Art and Cultural Materials"(Grant Number: S0991041), and MEXT Grant-in-Aid for Young Scientists (B) "Research on Information Access across Languages, Periods, and Cultures" (Leader: Akira Maeda, Grant Number: 21700271).

REFERENCES

1. Batjargal, B.; Khaltarkhuu, G.; Kimura, F & Maeda, A. (2010a). An Ancient-to-modern Information Retrieval for Digital Collections of Traditional Mongolian Script. *Proceedings of the 12th International Conference on Asia-Pacific Digital Libraries (ICADL2010)*, pp. 25–28, ISBN 978-3-642-13653-5, Gold Coast, Australia, June 2010, Springer-Verlag, Berlin Heidelberg

2. Batjargal, B.; Khaltarkhuu, G.; Kimura, F & Maeda, A. (2010b). An Approach to Ancient-to- modern and Cross-script Information Access for Traditional Mongolian Historical Collections. *Conference Abstracts of Digital Humanities 2010*, pp. 279–282, ISBN 978-0- 9565793-0-0, London, UK, July 2010, Centre for Computing in the Humanities, King's College London
3. Batjargal, B.; Kimura, F. & Maeda, A. (2010c). Approach to Cross-Language Retrieval for Japanese Traditional Fine Art: Ukiyo-e Database. *Proceedings of the 14th European Conference on Research and Advanced Technology for Digital Libraries (ECDL2010)*, pp. 518-521, ISBN 978-3-642-15463-8, Glasgow, UK, September 2010, Springer-Verlag, Berlin Heidelberg
4. Chan, L. M. & Zeng, M. L. (2006). Metadata interoperability and standardization - A study of methodology, Part I: Achieving interoperability at the schema level. *D-Lib Magazine*, vol. 12, no. 6, (June 2006), ISSN 1082-9873
5. Choimaa, S. & Shagdarsuren, T. (2002*). Qad-un űndűsűn quriyangγui altan tobči, (Textological Study)*, Volume I, Centre for Mongol Studies, National University of Mongolia, ISBN 99929-0-119-5, Ulaanbaatar (in Mongolian)
6. Ernst-Gerlach, A. & Fuhr, N. (2007). Retrieval in text collections with historic spelling using linguistic and spelling variants. *Proceedings of the 7th ACM/IEEE Joint Conference on Digital Libraries,* pp. 333–341, ISBN 978-1-59593-644-8, Vancouver, British Columbia, Canada, June 2007, ACM , New York, USA
7. Gotscharek, A.; Neumann, A.; Reffle, U.; Ringlstetter, C. & Schulz, K. (2009). Enabling information retrieval on historical document collections: the role of matching procedures and special lexica. *Proceedings of the 3rd Workshop on Analytics for Noisy Unstructured Text Data*, pp. 69–76, ISBN 978-1-60558-496-6, Barcelona, Spain, July 2009, ACM, New York, USA
8. Hauser, A.; Heller, M.; Leiss, E.; Schulz, K. & Wanzeck, C. (2007) Information Access to Historical Documents from the Early New High German Period. *Proceedings of the 20th International Joint Conference on Artificial Intelligence*, Hyderabad, India, January 2007
9. Kawashima, M.; Akama, R.; Yano, K.; Hachimura, K. & Inaba, M. (2009). *New Directions in Digital Humanities for Japanese Arts and Cultures*. Nakanishiya Shuppan, ISBN 978-4- 7795-0324-5, Kyoto

10. Khaltarkhuu, G. & Maeda, A. (2006). Retrieval Technique with the Modern Mongolian Query on Traditional Mongolian Text. *Proceedings of the 9th International Conference on Asian Digital Libraries (ICADL2006)*, pp. 478–481, ISBN: 3-540-49375-1, Kyoto, Japan, November 2006, Springer-Verlag, Berlin Heidelberg

11. Khaltarkhuu, G. & Maeda, A. (2007). Building a Digital Library of Traditional Mongolian Historical Documents. *Proceedings of the 7th ACM/IEEE Joint Conference on Digital Libraries (JCDL2007)*, p. 483, ISBN 978-1-59593-644-8, Vancouver, Canada, June 2007, Springer-Verlag, Berlin Heidelberg

12. Khaltarkhuu, G. & Maeda, A. (2008). Developing a Traditional Mongolian Script Digital Library. *Proceedings of the 11th International Conference on Asia-Pacific Digital Libraries: Universal and Ubiquitous Access to Information*, LNCS, vol. 5362, pp. 41–50, ISBN 978-3-540-89532-9, Bali, Indonesia, December 2008, Springer-Verlag, Berlin Heidelberg

13. Kimura, F. & Maeda, A. (2009). An Approach to Information Access and Knowledge Discovery from Historical Documents. *Conference Abstracts of the Digital Humanities 2009(DH09)*, pp. 359–361, ISBN 978-061-52-9929-7, College Park, MD, June 2009, Maryland Institute for Technology in the Humanities

14. Kimura, F.; Toba, T.; Tezuka, T. & Maeda, A. (2008). Federated Searching System for Humanities Databases Using Automatic Metadata Mapping. *Proceedings of 9th International Conference on Dublin Core and Metadata Applications*, pp. 139–140, ISSN 1939-1366, Seoul, Korea, October 2009, the Dublin Core Metadata Initiative

15. Kitamura, M. & Matsumoto, Y. (1996). Automatic Extraction of Word Sequence Correspondences in Parallel Corpora, Proceedings of the 4th Workshop on Very Large Corpora, pp. 79–87

16. Koolen, M.; Adriaans, F.; Kamps, J. & Rijke, M. (2006). A Cross-Language Approach to Historic Document Retrieval. *Proceedings of the 28th European Conference on IR Research: Advances in Information Retrieval*, LNCS, vol. 3936, pp. 407–419, ISBN 978- 3-540-33347-0, London, UK, April 2006, Springer-Verlag, Berlin Heidelberg

17. Pilz, T.; Ernst-Gerlach, A.; Kempken, S.; Rayson, P. & Archer, D. (2008). The Identification of Spelling Variants in English and German Historical Texts: Manual or Automatic. *Literary and*

Linguistic Computing, vol. 23, no. 1, November 2008, pp. 65–72, ALLC, ACH, Oxford University Press, ISSN 0268-1145, Oxford, UK

18. Shagdarsuren, T. (2001). *Study of Mongolian Scripts (Graphic Study of Grammatology)*, Enlarged second edition. ISBN 99929-5-347-0, Urlakh Erdem Khevleliin Gazar, Ulan Bator (in Mongolian)
19. Tanaka, T. (2002). Measuring the Similarity between Compound Nouns in Different Languages Using Non-Parallel Corpora, Proceedings of the 19th COLING, pp. 981– 987, ISBN 978-1-55860-896-2, Morgan Kaufmann Publishers, December 2002, Taipei, Taiwan
20. The Dublin Core Metadata Initiative (2008). Dublin Core Metadata Element Set, Version 1.1. http://dublincore.org/documents/dces/
21. Tsevel, Y. (1966). *Mongol helnii tovch tailbar toli.* Ulsiin Khevleliin khereg erkhlekh Khoroo, Ulan Bator (in Mongolian)
22. Tungalag, D. (2005). *Mongol ulsiin undesnii nomiin san dahi Mongoliin tuuhiin gar bichmeliin nomzuin sudalgaa*, Volume 1. ISBN 99929-6-313-1, Time Printing, Ulan Bator (in Mongolian)
23. Zeng, M. L. & Jian, Q. (2008). *Metadata*, ISBN 978-1-85604-655-8, Facet Publishing, London

13
CHAPTER

MULTITASKING MADE EASY: SUPPORTING ACADEMIC WRITING IN DIGITAL LIBRARIES WITH AN AMBIENT SEARCH SYSTEM

Anatoliy Gruzd[1] and Michael B Twidale[2]

[1] *Dalhousie University, Canada*

[2] *University of Illinois at Urbana-Champaign, USA*

INTRODUCTION

When personal computers first arrived, many predicted that by the end of the century our desktops would become free from clutter as we moved to the paperless office. However, if we look at desktops of current researchers, we can see that this has definitely not happened (see Figure 1). But it is not only researchers' physical desktops that are cluttered; their computer desktops (and file systems) are just as cluttered. This should not come as a surprise. Because the process of writing is such a complex process, people need this space to lay out their notes, readings, drafts, printouts, etc. As a result, the desktop is literally disappearing under piles of paper, just not in the way predicted.

Writing is an extremely complex process, consisting of multiple components that can occur in many different combinations. With respect to academic writing, these components include planning and organization, presentation of facts, marshalling of arguments, preparing tables, figures, diagrams and references, searching for references, reading, annotating related and possibly related work, citation chaining, countering argument threads in related work, searching for inspiration, and searching for supporting and refuting evidence in the literature (Brockman et al., 2001; Fister, 1992; Kuhlthau, 2004; Palmer, 2005; Torrance et al., 1994; Wolber et al., 2002). At times it can help to consider these activities, some of which are very different, as a special kind of multitasking, raising the question of how best to support it.

Figure 1. Example of a cluttered desktop and office of a person (one of the authors) engaged in research and academic writing.

Before personal computers, writing involved the accumulation of resources from books, journals, preprints, the authors' prior work, notebooks, outlines and previous drafts. These paper sources might be laid out on available flat surfaces (desks, tables, often the floor) in order to be easily available during writing, not only for immediate reference, but to serve as passive reminders and awareness agents of other activities to do and issues to consider. Working in an office means that even vertical surfaces can be recruited - whiteboards containing the results of discussions with co-authors, and bookshelves serving as reminders of other resources that might be considered, or as sources of inspiration when writing meets an impasse.

What is so noticeable is that since the arrival of personal computers in the mid 1980s, and despite the enormous improvements in processing power, memory capacity and screen size, as well at the dramatic improvements in online access to digital libraries, web pages, scholarly repositories etc., the paper-strewn scenario of the previous paragraph is still all too familiar. If anything, access to more resources more quickly just means that it is easier to print out and strew more documents around one's office while writing. Additionally, the access to online full text versions of papers means that researchers can also accumulate substantial personal digital collections of papers on their hard drives. If poorly organized, these collections themselves are difficult to navigate, and are in need of a context-aware retrieval system (Kljun & Carr, 2005).

The issue in interface design is that in the context of writing the desktop metaphor is *not* necessarily bad or obsolete. Rather the problem is that it has never really been tried. Desktop computers have limited screen real estate and so must employ various tricks of overlapping, iconization, listing of files, links and other tricks to manage the quantity and diversity of materials needed to support writing. Even with 24" screens and double, even triple monitors, the available space is small compared to most people's actual desktops (let alone their floors and walls that can be employed in the more intense periods of writing). If anything, current screens resemble the tiny individual desktops used by examination candidates (typically less than 3' by 3'), with just enough room for a question paper and an answer booklet (see Figure 2). That is fine for a test of memory, but is wildly unsuitable for rich, contextualized writing, let alone thinking and searching while writing.

The growing availability of academic literature on the Internet makes it possible for authors to multitask by switching between writing and information searching; as easy as switching between two windows on their computer monitors. However, this poses a new challenge to writers. The complexity of information searching (generating query, browsing, assessing results, etc) makes it very hard to stay focused on writing. Furthermore, an overwhelming diversity of poorly connected online research tools (search engines, digital libraries, bookmarking sites, etc) often leads to information fragmentation among their users (Boardman & Sasse, 2003).

Retrieved from
http://www.flickr.com/photos/comedynose/3571102858

Figure 2. The existing manifestation of the desktop metaphor implies a very tiny desktop, more like that used for examinations involving writing from memory than larger office desktops (plus floors and walls) used for writing from resources.

We are exploring the design of an online context-aware retrieval system that allows users to multitask by writing while still actively engaged in research activities online. We have developed a web-based prototype called PIRA (Personal Information Research Assistant) available at http://writeNcite.com. PIRA relies on external digital libraries and search engines to produce a list of academic references related to what a user is writing at the moment; thus, allowing authors to stay focused on their writing.

PIRA attempts to support many of the awareness features provided by a papers strewn desktop and floor, but with a more active approach, providing an ambient awareness of work that *may* be relevant and inspirational. In its current form, PIRA remains locked into a single screen, although attempting to use it in a more integrated manner for supporting writing.

This paper focuses on the methodological aspects of user-centric evaluations of PIRA. More specifically, we investigate influences of multitasking as supported by PIRA on users' writing.

RELATED WORK

Previous research in this area produced a number of interesting design solutions for context- aware retrieval systems that can support writing and reading. Some examples of earlier work in this area include: Watson to support desktop-based writing/reading activities (Budzik et al., 2002), Implicit Queries (IQ) to support composing/reading e-mails (Dumais et al., 2004), Phrasier to support interactive document retrieval using keyphrases (Jones & Staveley, 1999), and systems like Letizia and PowerScout (Lieberman et al, 2001) and WebTop (Wolber et al., 2002) to support browsing and reading of web pages. There are also some context-aware desktop search engines for local files like Remembrance Agent, Margin Notes, and Jimminy (Rhodes, 2003). And more recently, there are also the Context Creation Tool which is designed to support interactive reference gathering, academic note taking and writing (Berendt et al, 2010) , CONTEXT – a context-aware information retrieval system for bloggers (Gruzd & Wong, 2010), and the Context Awareness Tool (CAT) - a general-purpose "writing for the web" tool (Powell et al., 2009).

The existence of these tools show that there is both interest and progress being made to develop a truly context-aware retrieval system that can better support users' information behavior. However, there is still a lot more work that needs to be done to address all of the challenges associated with writing (especially, academic writing) in the context of using digital libraries. Some of the challenges include designing interfaces that are more context- aware (e.g., Ruthven, 2008) and interfaces that enable users to serendipitously discover new ideas (e.g., Toms & McCay-Peet, 2009). One system that has been designed to address some of these challenges is PIRA.

PIRA'S USER INTERFACE

In this section, we provide a brief overview of PIRA. A more detailed description of PIRA and review of the related work can be found elsewhere (Gruzd & Twidale, 2006; Twidale et al., 2008).

PIRA is a web-based writing tool with two main interface components. On the left is a basic text editor. On the right is an area for searching and managing references. As the user writes (or pastes) new content in the editor, PIRA automatically extracts significant search keywords, displays them in the "Suggested Search Terms" pane, and retrieves and presents suggested references in the "Auto References" pane (see Figure 3). These can be simply ignored if the user is focusing on writing, or glances at them and deems them irrelevant. They are gradually replaced by alternate suggestions based on the current area of writing activity in the text editor. Mousing over a reference provides more details including the paper's abstract. If a user sees a potentially useful citation among the sources suggested by

PIRA, they can temporarily lock the citation by clicking on the Pin icon in the front of the suggested reference. Along with temporarily locking the citation, a user can also open the full-text of an article in a new window (if provided by the digital library) or save the citation for use in future sessions using the Disc icon. If a user decides to save a citation for later use, he or she will have the option of associating the suggested reference with the current document or with any other (previously saved) documents. This is done to help users avoid situations when they "often don't remember that they've already saved potentially useful or meaningful material" (Fister, 1992). Once a citation is saved, it can be later accessed using a built-in bibliographic management interface (see Figure 4).

PIRA automatically and continuously suggests new reference sources related to the content of users' writing. Reference sources are suggested from various open-access digital libraries and search engines. By default, Google.com and CiteUlike.org are automatically selected. Users can leave the default selection or build their own personal list. For example, depending on their research topic, users may prefer to only use specialized data sources like CiteSeer.IST (Information and Computer Science – oriented), or they might choose to use only general sources like the Directory of Open Access Journals (DOAJ) (http://www.doaj.org). To select or deselect an external data source, users just simply need to click the "USE" or "DON'T USE" button accordingly (see Figure 5).

PIRA has been built as a web mashup, using a variety of different web services (different text editors, text and concept extraction services, and different user-specifiable digital libraries) enabling us to undertake a systematic exploration of the design space of variants on writing support and ambient search.

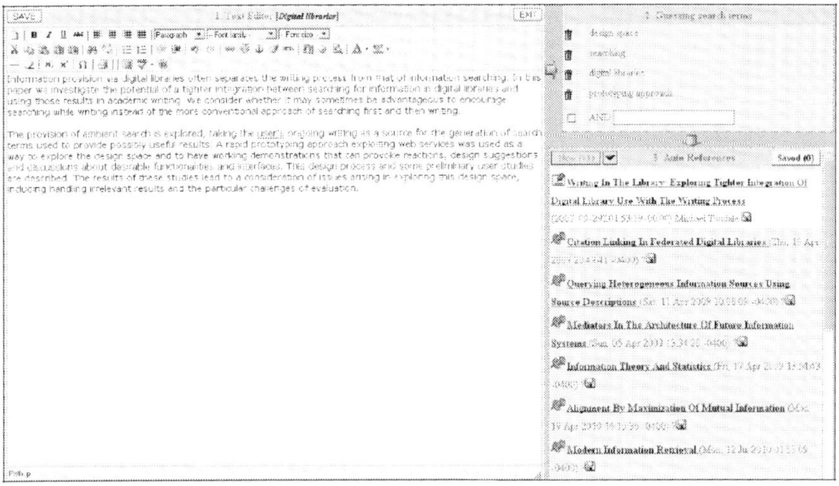

Figure 3. PIRA's main display showing integration of writing and ambient searching.

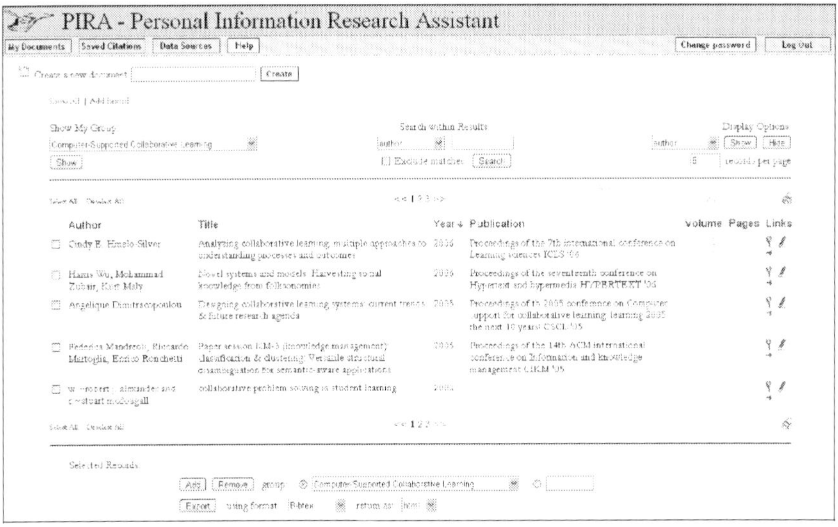

Figure 4. Built-in bibliographic management interface based on RefBase, open source software.

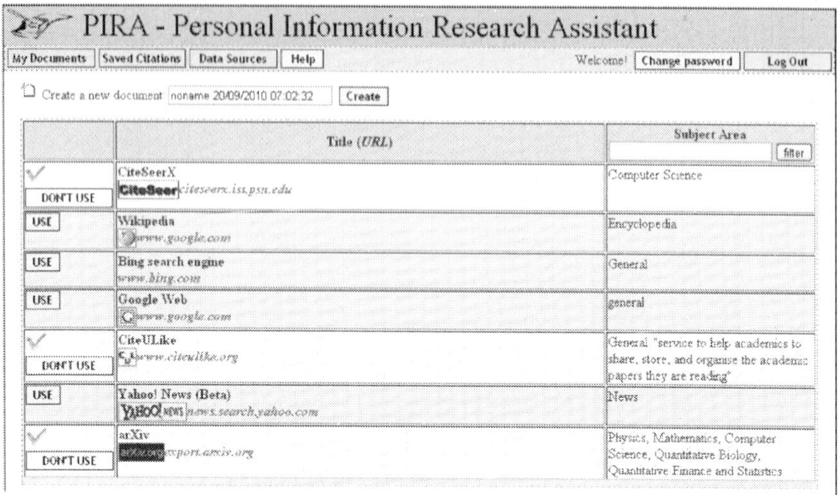

Figure 5. Reference sources that are available in PIRA.

USER STUDY DESIGN

Research questions and method

In this chapter we try to address two research questions:

- Is multitasking in the context of writing and searching (as supported by PIRA) a manageable process that might be adopted and accepted by the average user?
- Is multitasking (as supported by PIRA) more effective than the more traditional approach to academic writing (first searching and then writing)?

To answer the first question, we need to identify whether users were actually engaged in the process of writing and searching in a sequential manner (one activity followed by the other) or in a more parallel manner (working on both simultaneously or frequently switching between the two). We will refer to the parallel manner as multitasking or a Write While You Search (WWYS) approach. In our analysis, we look for instances when a user was switching back and forth between writing and searching activities. We note that a user applied the WWYS approach whenever he or she consulted (e.g. accessed or saved) references suggested by PIRA in between making any changes to the text. Or, better yet, when a user actually cited any of references suggested by PIRA.

To answer the second question, we need to operationalize the effectiveness of PIRA usage. One way to do this is to conduct an assessment of reference gathering productivity. There are two parameters that can be used for this purpose: the number of references accessed by the user and the number of references saved by the user. The higher value of either of these parameters per user per session may be characterized as evidence of a higher level of effectiveness of PIRA usage. The reason we need to examine both parameters is because a user may access a relevant reference without saving it into his or her online account, or vice versa. Finally, we compare average effectiveness between users using the WWYS approach and those using the sequential model. To answer both questions, we rely on content analysis of each user's drafts, quantitative analysis of users' interaction logs and users' responses to an online questionnaire.

Users and data collection

All participants in our study were volunteers invited at our demo sessions or by their colleagues or professors who attended our demo sessions. To become a user, a person has to create an online account with PIRA and agree to the terms of the informed consent. Participants were asked to use PIRA for their research-related writing tasks. Users could create and modify any text documents and reference collections. They could discontinue their participation at any time for any reason.

PIRA recorded users' writing, what it recommended, and what if anything users chose to do with those recommendations. After the user has used PIRA for three times, the user was asked some questions via an online questionnaire about what he or she thought about the use of PIRA, and how it might be improved. Completing this questionnaire was optional.

The log data was collected for the period of four months. During this period, our users consisted of two main groups: 11 undergraduate students in the English program and 14 graduate students in the Library and Information Science (LIS) program. On average, users in each group visited PIRA a similar number of times; 7 and 5 for undergraduate and graduate users respectively.

Data analysis

We began the analysis by identifying common approaches to writing used in PIRA by examining the content of papers and how that content changed over time. We distinguished four main writing approaches:

- **Keywords**. This approach is very similar to the way people search on the Internet. The user starts with a list of main keywords/concepts to describe a problem domain, and then modifies this list depending on the retrieved results.
- **Freewriting**. The user writes down full sentences about issues they want to address and/or statements about their prior knowledge of these issues.
- **Copy & Paste**. The user copies a chunk of pre-written text to PIRA's text editor. The chunk may range from a short paragraph to several pages.
- **Think by Writing**. The user writes down his or her thoughts when assessing the relevance of suggested references.
- In addition to four common writing approaches described above, we added one new approach for the graduate students - *Proof Reading*. This approach involves only minor changes to the text, mostly to make one's writing sound better.

Tables 1 & 2 below show which approaches and in what order were used by each user. To indicate whether or not the user switches between writing and searching activities (the WWYS approach) or only was focusing on one activity at the time, we use an additional category: *Monitoring References*. If the user is engaged in *Monitoring References* this means that he or she completely stopped writing (at some point) and exclusively began interacting with some suggested search keywords and references.

Although *Copy & Paste* can be conducted in parallel with searching, we will not consider it as an indicator of the WWYS approach. This is primary because the copied text was already written prior to the session with PIRA. Similarly, we will not consider the use of the *Keywords* only approach as WWYS because this approach does not produce any coherent text that can be used in the future paper. However, if a user started with *Copy & Paste* or *Keywords*, but then switched to *Freewriting*, and he or she was not engaged in *Monitoring References*, then the overall user's approach can be considered as WWYS.

The following section presents results for each group separately. To ensure confidentiality, we refer to each user by a code number.

Table 1. Common writing approaches in PIRA for **undergraduate** students (each cell value indicates the order in which the particular approach occurred).

Approach \ User ID	Keywrods	Freewriting	Copy & Paste	Think by Writing	Monitor References
49	#1	#2			
50		#1			
51		#1			
52		#1			
53	#2			#1	
54		#1			
58		#1			
59		#1			#2
60		#1			
67		#2	#1		
71			#1		#2

RESULTS

1.1 Undergraduate students

Q1. Is multitasking in the context of writing and searching (as supported by PIRA) a manageable process that might be adopted and accepted by the average undergraduate user? Based on the content analysis, we can conclude that the majority of undergraduate students in the study preferred to use *Freewriting* as their primary writing approach. This preference may be influenced by the fact that they are all English majors. The log data shows that as they wrote, the users were also interacting with various search features of the system such as accessing and saving relevant references. Different users were more or less involved in the use of these features. Although the majority of users accessed and read potentially relevant references as soon as they noticed them; there were two users (ID# 58 and 60) who preferred not to stop writing when they saw something interesting. Instead, they simply saved any potential relevant references to read later. (Such behavior is characterized by the much higher number of saved references than those that were accessed while in the midst of writing.) Despite these differences in reactions to potential relevant references, we can consider both types of behaviors as multitasking to some degree since both involved writing and reference assessment/gathering.

Table 2. Common writing approaches in PIRA for **graduate** students (each cell value indicates the order in which the particular approach occurred)

Approach \ User ID	Keywords	Freewriting	Copy & Paste	Think by Writing	Proof Reading	Monitor References
28			#1		#2	
31			#1			#2
32		#1				
33		#1				
37*	#1					#2
41		#1	#2			#2
44*		#1				#2
55	#1					#2
57*			#1			#2
62*			#1			#2
65	#1	#3		#2		
69*	#1					#2
84*		#1				#2
92			#1			#2

Additional evidence that multitasking as supported by PIRA was manageable can be found in the responses to the online survey. Many undergraduates expressed their support of the idea of multitasking and switching between writing and searching. As one student noticed, nowadays multitasking is part of their normal behavior on the Internet. They often do instant messaging with their online friends, browser websites, work on a class assignment, search for information, etc, simultaneously. This is a more extreme form of multitasking than the one we are referring to in this chapter – the switching between different components of the overall single goal of writing.

In general, our analysis of the current data suggests that the majority of the undergraduate users were able to successfully adopt the WWYS approach.

Q2. Is multitasking (as supported by PIRA) more effective than the more traditional approach to the academic writing (searching and then writing)?

The fact that two users, who were not engaged in the WWYS approach, accessed and saved a significantly smaller number of references compared to the group's average, may suggest that the WWYS approach is *likely* to be more productive than a more traditional approach. Unfortunately, since the majority of users preferred the WWYS approach over the other, we do not have enough data to draw a conclusion as to which of two approaches is more productive. Further testing is necessary to answer this question.

Graduate students

Q1. Is multitasking in the context of writing and searching (as supported by PIRA) a manageable process that might be adopted and accepted by the average graduate student user?

Graduate students did not multitask much (except 4 users). About half of all graduate students in the study re-used parts of their papers written elsewhere and simply pasted them into PIRA. Many other users either wrote one or two sentences stating their research topic or simply typed a few search keywords into the text editor. After this, they were mostly monitoring references. This type of behavior suggests that graduate students in our study perceived PIRA primary as a searching tool rather than a writing tool. This does not necessarily mean that multitasking was not manageable, but rather that users adopted other ways of using PIRA (to our pleasant surprise) which will be one of the subjects of our future evaluations. As with other context-aware interfaces, it is important that developers be reminded that a measure of success is not merely that the application is adopted by those users who work in the way that the developers intend, but that the application is flexible enough that alternate, even novel uses are facilitated or at least not impeded.

Q2. Is multitasking (as supported by PIRA) more effective than the more traditional approach to the academic writing (searching and then writing)?

To conduct a reliable comparison, we decided to exclude the 6 "spectators"-users from this analysis (marked with an asterisk * in Table 2). This is due to the fact that their interactions with PIRA were limited to only few minutes. As a result, it is hard to say which approach they really followed or would follow. (After a close examination of the log data and documents created by these 6 users, we came to the conclusion that these users were just checking out PIRA's functionalities.)

Among the 8 remaining users, there were 4 users (ID# 28, 32, 33, 65) who were engaged in multitasking (further referred to as Group A) and 4 users (ID# 31, 41, 55, 92) who were not (Group B). Group A accessed an average of 20 and saved an average of 13 references per user. Group B accessed an average of 8 and saved an average of only 4 references per user. In sum, users who were engaged in multitasking accessed about 2 times more and saved 3 times more references per user than those who were not. Therefore, although based on our very small sample, we would claim that in general, multitasking seems to be a more productive approach, despite the risks of distraction.

CONCLUSIONS AND FUTURE CONSIDERATION

The results from our small scale user study suggest that undergraduate users are more likely to multitask between writing and search-related activities than graduate users. The difference between two groups may be due to the difference in their tasks. Most graduate users were focusing on finding relevant references (Keywords approach) or making sure that they had already cited all relevant references (Copy & Paste approach); whereas undergraduate users were more concerned with producing coherent text. This, in turn, may be explained by many different factors such as specific requirements of the assignment, students' major and/or upcoming deadlines. However, this matter requires further investigation.

Our second conclusion is that in general those users from both groups who did use the multitasking approach demonstrated a more productive reference gathering behavior than those who did not. This can be explained from the system's point of view. Users who multitasked were often modifying their drafts. As a result, PIRA was able to suggest references that are related to newly emerged themes in the text. Furthermore, users who multitasked also interacted with suggested search keywords and references more frequently than those who did not. Since every such interaction provided PIRA with user's relevance feedback, the system was able to significantly improve the relevance of its suggestions PIRA enabled users to manage the information that could inform their evolving thinking, supporting that fine line between an overly narrow focus just on the resources currently on the users' mind, and the insights and balance that come from a larger perspective, but one that can lead to endless distraction.

In our future work on PIRA, we are planning to increase the size of our user sample as well as diversify its population by including students and faculty members from disciplines other than English and LIS. Also we are planning to explore other ways to measure the effectiveness of PIRA usage to consider the quality of gathered references as well as their impact on the completion of the user-specific tasks (for example, to complete a literature review versus prepare a paper outline versus come up with creative ideas).

The challenge of context-aware retrieval for supporting the web-based writing process is similar to that of the more conventional issues of planning, coordinating appointments and organizing and accessing personal files. However, it has rather different emphases. The fluid, rapid multitasking nature of the different components of web-based writing means that it is important to support low-effort context-aware information retrieval. There is also a need to support both awareness of the familiar (papers already read, issues that must be considered), as well as the unfamiliar (new papers not yet considered, possibly relevant related work, interesting inspirational insights etc.). As with much research on Digital Libraries, there is a recurrent need to help people handle complexity and diversity, and to exploit context and visibility as a way of supporting ambient rather than distractingly intrusive awareness. Finally the diversity of ways in which people accomplish complex tasks such as writing reminds us that digital libraries must accommodate a wide range of use patterns rather than forcing users to conform to an idealized mode of acting. Such diversity typically means allowing for a range of appropriation activities including combination with other resources and tailoring to fit both long term user preferences and the particular needs of the current task at hand.

ACKNOWLEDGEMENTS

This work was partially supported by the Social Sciences and Humanities Research Council (SSHRC) grant.

REFERENCES

1. Berendt, B., Krause, B., Kolbe-Nusser, S. (2010). Intelligent scientific authoring tools: Interactive data mining for constructive uses of citation networks. *Information Processing & Management*, 46(1), 1-10.
2. Boardman, R. and Sasse, M.A. (2003). Too Many Hierarchies? The Daily Struggle for Control of the Work-space. *HCI International'03 International Conference on Human-Computer Interaction*, Crete, Greece.
3. Brockman, W.S., Newmann, L., Palmer, C.L., and Tidline, T.J., (2001). *Scholarly Work in the Humanities and the Evolving Information Environment*, Washington, DC: Digital Library Federation and Council on Library and Information Resources.
4. Budzik, J., Bradshaw, S., Fu, X. and Hammond, K. (2002). Supporting online resource discovery in the context of ongoing tasks with proactive software assistants. *International Journal of Human–Computer Studies*, 56 (1), 47–74.
5. Dumais, S., Cutrell, E., Sarin, R. and Horvitz, E. (2004). Implicit queries (IQ) for contextualized search. *Proceedings of the 27th Annual International Conference on Research and Development in Information Retrieval*, ACM Press, New York, NY, pp. 594-594.
6. Fister, B. (1992). The Research Processes of Undergraduate Students. *Journal of Academic Librarianship*, 18 (3), 163-169.
7. Gruzd, A.A. & Twidale, M.B. (2006). Write While You Search:Ambient Searching of a Digital Library in the Context of Writing. *Proceedings of the 1st International Workshop on Digital Libraries in the Context of Users' Broader Activities (DL-CUBA) at the Joint Conference on Digital Libraries(JCDL'06)*, Chapel Hill, NC, USA, pp. 13-16.
8. Gruzd, A. and Wong, J. (2010). Blogging with CONTEXT: A context-aware information retrieval system for bloggers. *Proceedings of the 3rd Information Interaction in Context Symposium*, ACM Press, New York, NY, pp 289-292.

9. Jones, S. and Staveley, M.S. (1999). Phrasier: a system for interactive document retrieval using keyphrases. *Proceedings of the 22nd Annual International Conference on Research and Development in Information Retrieval*, ACM Press, New York, NY, pp. 160–167.
10. Kljun, M. and Carr, D. (2005). Piles of Thumbnails - Visualizing Document Management. *Proceedings of the 27th International Conference on Information Technology Interfaces (ITI2005)*, Cavtat, Croatia.
11. Kuhlthau, C.C. (2004). *Seeking Meaning: A Process Approach to Library and Information Services*. 2nd ed., Westport, CT: Libraries Unlimited.
12. Lieberman, H., Fry C. and Weitzman, L. (2001). Exploring the web with reconnaissance agents. *Communications of the ACM*, 44 (8), 69–75.
13. Marshall, C.C. and Jones, W., (2006). Keeping Encountered Information. Commun. ACM. *Communications of the ACM*, 49(1), 66-67.
14. Palmer, C.L. (2005). Scholarly Work and the Shaping of Digital Access. *Journal of the American Society for Information Science and Technology*, 56(11), 1140-1153.
15. Powell, J.E., Collins, L.M., Martinez, M.L.B. (2009). The Fierce Urgency of Now: A Proactive, Pervasive Content Awareness Tool. *D-Lib Magazine*, 15 (5/6).
16. Rhodes, B. (2003). Using physical context for just-in-time information retrieval. *IEEE Transactions on Computers*, 52 (8), 1011–1014.
17. Ruthven, I. (2008). The context of the interface. *Proceedings of the Second International Symposium on Information Interaction in Context*, London, United Kingdom, pp. 3-5.
18. Toms, E.G. and McCay-Peet, L. (2009). Chance Encounters in the Digital Library. In Agosti, M., Borbinha, J., Kapidakis, S., Papatheodorou, C., Tsakonas, G. (Eds.). *Research and Advanced Technology for Digital Libraries*, Springer: Berlin / Heidelberg, 192-202.
19. Torrance, M., Thomas, G. V., & Robinson, E. J. (1994). The Writing Strategies of Graduate Research Students in the Social Sciences. *Higher Education*, 27 (3), 379-392.

20. Torrance, M., Thomas, G.V., and Robinson, E.J. (1994). The Writing Strategies of Graduate Research Students in the Social Sciences. *Higher Education*, 27 (3), 379-392.
21. Twidale, M.B., Gruzd, A.A. & Nichols, D.M. (2008). Writing in the Library: exploring tighter integration of digital library use with the writing process. *Information Processing & Management*, 44(2), 558-580.
22. Valentine, B. (1993). Undergraduate Research Behavior: Using Focus Groups to Generate Theory. *Journal of Academic Librarianship*, 19(5), 300.
23. Wolber, D., Kepe, M. and Ranitovic, I. (2002). Exposing document context in the personal web. *Proceedings of the 7th International Conference on Intelligent User Interfaces*, ACM Press, New York, NY, pp. 151–158.

14

CHAPTDR

BRINGING THE DIGITAL LIBRARY DESIGN INTO THE REALM OF ENTERPRISE ARCHITECTURE

A Abrizah and A.N. Zainab

Digital Library Research Group, University of Malaya, Kuala Lumpur, Malaysia

INTRODUCTION

Previous digital library research and initiatives have conceptualized and proposed several frameworks for the design, development, evaluation and interaction of digital library systems. Levy and Marshall (1996) discussed a work-oriented perspective of digital library research that is based on the work people do, and how digital libraries assist in the completion of work related tasks. Their framework highlights three crucial characteristics of digital libraries: document, technology and work (which involves research and service). Moen and McClure (1997) identified a framework of five interacting dimensions in digital library of Government Information Locator Service (GILS): policy, users, technology, contents, and standards. The evaluation framework also includes three perspectives, representing the "views" of the stakeholders in the GILS: users, agencies, and the government. Marchionini and Fox (1999) identified four dimensions of digital library development: community, technology, service and content. Saracevic and Covi (2000) presented a framework, consisting seven levels, for examining digital libraries: social, institutional, individual, interface, engineering, processing and content. Another holistic framework is presented by Fuhr et al. (2001) consisting four major dimensions, namely data/collection, system/technology, users and usage. Sandusky (2002) developed a list of six attributes in framing digital library usability research: audience, institution, access, content, services, and design and development. Soergel (2002) offered a digital library research framework consisting of three guiding principles and eleven specific themes for research and development. Gonclaves et al. (2004) introduced 5S and formalisms for Streams, Structures, Spaces, Scenarios, and Societies — as a framework for providing theoretical and practical unification of digital libraries. All these frameworks emphasize the importance of a holistic approach rather than examine digital libraries as a single view, which would be limited in their utility.

However, the absence of common frameworks in the digital library development practices undermines the ability to develop and design digital library systems efficiently, to create large-scale collaborative activities, and to communicate the value of the systems to other communities. Gladney et al. (1994) wrote that the broad and deep requirements of digital libraries demand new frameworks and theories in order to understand better the complex interactions among their components. Supporting this claim, the summary report of the Joint NSF-European Union (EU) Working Groups on Future Directions of Digital 1 Libraries Research recommended that "new frameworks and theories be developed in order to understand the complex interactions between the various components in a globally distributed digital library" (Schauble and Smeaton, 1999). Formal frameworks are crucial to specify and understand clearly and unambiguously the characteristics, structure, and behavior of complex information systems such as digital libraries. The Digital Library Federation (DLF) in 2005 sponsored the formation of the Service Framework Group (SFG) to consider a more systematic, community-based approach to align the functions of digital libraries in fulfilling the needs of information environments (Lavoie, Henry and Dempsey, 2006). DLF envisaged that digital library functionalities be generated from the library business processes, considering the architectures as information systems with specific business requirements (Castelli and Fox, 2007). It is in this context, and in recognition of visions already underway to align digital library development with the emerging perspective of the Enterprise Architecture (Borninha, 2007), the authors conducted a study that seeks to understand and model the digital library services adopting a framework that give preference to scopes, goal requirements and processes – those concepts already common in Enterprise Architecture processes

Abdullah and Zainab (2008) regard a digital library as an enterprise that requires architecting. An Enterprise Architecture for the digital library is a framework or blueprint which shows how the digital library organisation carries out an intended task and how the digital library will or can improve the processes. It shows how a digital library represents a special workspace for the user community, not only for search and access but also for the process or workflow management, information creation, sharing and exchange, and distributed workgroup communication. In order to identify what is required of a digital library in a specific context, a sound methodology is needed to establish an understanding of the digital library entire structure. A multi-faceted information services such as digital libraries may be examined along different dimensions and from different perspectives or views of the stakeholders. There is a need to identify potential users, their involvement and roles in the digital library, their attitude towards the technology, their perception of its potential use and how it fits within the digital library goals in general. In order to do this, a digital library enterprise is required, which is derived and based on empirical data and stand up to conceptual reasoning. This chapter shows how an established Enterprise Architecture can be adopted as a formal framework to guide the research, design and development of digital libraries, providing a precise specification of requirements against which the implementation can be compared for correctness.

ENTERPRISE ARCHITECTURE

In general, Enterprise Architecture is a framework that describes how an organisation develops, manages and uses information technology to optimally support its business functions (Kahn and Wilensky, 1995). Sometimes, the term refers to the group of people responsible for modeling and then documenting the information architecture; other times the term denotes the process of doing this work. More commonly, Enterprise Architecture refers to the models, documents and reusable items (such as components, framework and objects) that reflect the actual architecture (McGovern et al., 2003). In the EACommunity (http://www.eacommunity.com/) Enterprise Architecture is a framework or blueprint for how the organisation achieves the current and future business objectives. It examines business processes, information technology, software and hardware, local and wide area networks, people, operations and projects with an organisation's overall strategy. Each of these strategies has a separate architectural discipline (such as business, information, application, technical and product) and Enterprise Architecture is the "glue" that integrates each of these disciplines into a cohesive framework (Bolton, 2004) as depicted in Figure 1.

From these definitions, it is understood that Enterprise Architecture consists of the various structures and processes of an organisation. Following this understanding, it is known that an Enterprise Architecture model is a representation of those structures and processes. A good Enterprise Architecture model will depict the organisation both as it is today and as it is envisioned in the future, and will map the various views representing the architecture to one another. These views include both business-oriented perspectives as well as technical perspectives. The blueprint or framework of the enterprise would reveal detailed statements and processes that characterized architectural drawings. The detail drawings would be in any form, such as rich pictures, structured charts, data flow diagrams, Unified Modelling Language (UML) activity diagrams, database tables and entity-relationship model.

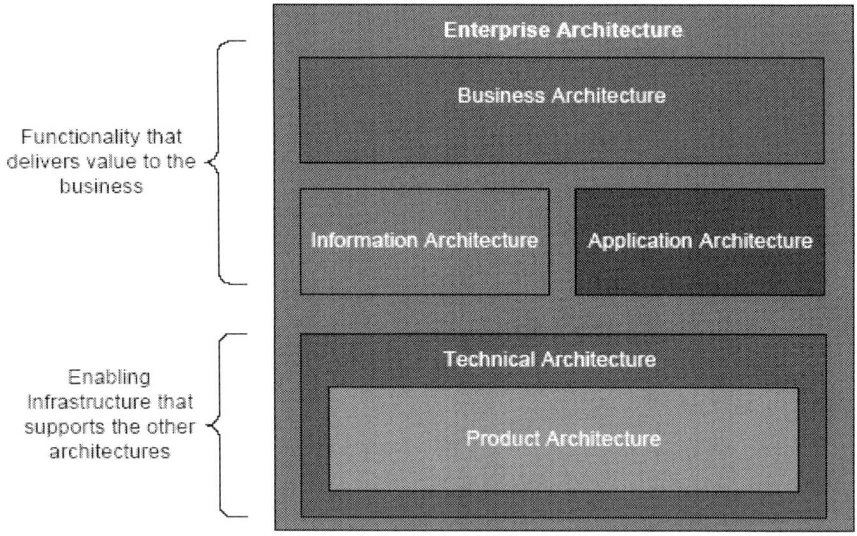

Figure 1. Enterprise Architecture Relationship (Source: Bolton, 2004)

ZACHMAN FRAMEWORK FOR ENTERPRISE ARCHITECTURE

The Zachman Framework is a logical structure for classifying and organising the descriptive representations of the enterprise that are significant to the management of the enterprise, as well as to the development of the enterprise' systems (Zachman, 2002). The framework uses a grid model to provide a logical structure for classifying and organising the descriptive representation of an enterprise, in six different dimensions, and each dimension can be perceived in five different perspectives. In this framework, the architecture is described across two independent aspects, the rows represent the views of five different types of stakeholders (planner, owner, designer, builder and sub-contractor) and the columns represent six different aspects of the architecture (data, function, network, people, time and motivation). The points of intersection between the rows and the columns (between the views and the aspects) form cells. Each of these cells holds important information of the enterprise (also known as artifacts) that needs to be understood and explicitly declared. The Zachman Framework's enterprise design model is presented in Figure 2.

The Zachman Framework for Enterprise Architecture is found suitable to investigate the initial requirements and define the digital library organisation, processes, technology and information flows, as well as ground the design of digital libraries for the following reasons:

a. The framework helps to explicitly show the many perspectives that need to be addressed by the digital library. It requires the planner, owner and designer of the digital library to involve the stakeholders to ensure that it meets their needs and will be used. It holistically controls the approach to investigate the user requirements and guides the data gathering techniques.

b. The framework requires the involvement of stakeholders, not just the enterprise architects and developers. and ideally this practice is what digital library designers and developers should follow. This aligns with the need to involve stakeholders in digital library design and development.

c. The framework is robust enough. It explicitly shows and requires the designer to consider all aspects (What, How, Where, Who, When, Why) of the digital library design.

d. The framework is generic in nature (Pereira and Sousa, 2004) and can be applied perfectly to digital library organization. As such it is a flexible framework and it does not impose a method or restrict any user to a set of pre-defined artifacts.

This chapter shows, through a case study, how the three tiers of Zachman Framework – the contextual (scope) or planner's perspectives (Figure 2 Row 1), the conceptual (business model) or the owner's perspectives (Figure 2 Row 2) and the logical (system model) or the designer's perspectives (Figure 2, Row 3) – are used to design a digital library. The first two rows or layers are referred to as the Business Architecture (Figure 1) for they describe the functions a business performs and the information it uses. The third row refers to the information and the application architecture (Figure 1). The planner is concerned with positioning the digital library in the context of its environment. This is when the planner enquires about the demographics of the stakeholders, ICT individual differences, their readiness to participate and collaborate, their awareness of the concept of digital libraries and their perception of the digital library initiative. The owner is interested in the digital library's deliverable and how it will be used. The designer is concerned with how the digital library is to perform its functions. This involves investigating the resources that are used, the user behaviour of seeking for resources, the experience of searching, the relevance perceived and the problems encountered. The possible sets of constructs or artifacts to represent the cell content for each cell in the top three rows or layers of the Zachman Framework are presented in Table 1.

Row 6 of the framework represents the physical manifestation of the end product itself. Zachman (2003) says that technically Row 6 is not an architecture because it is not a representation (it is the actual thing), however it is useful to incorporate it into the framework graphic as it completes the architectural picture. For an enterprise employment of the framework, Row 6 represents the Functioning Enterprise, which is the end result of the architectural process. In this research, the end result is to ensure that Row 6 represents what the owners have in mind for the digital library enterprise at Row 2. Therefore it is important to incorporate users assessment to evaluate the viability of a useful and enduring digital library system for the user community. This would involve assessment of the usefulness and usability of the system in terms of resources (data), processes (function), location (network), user community (people), time and goals (motivation).

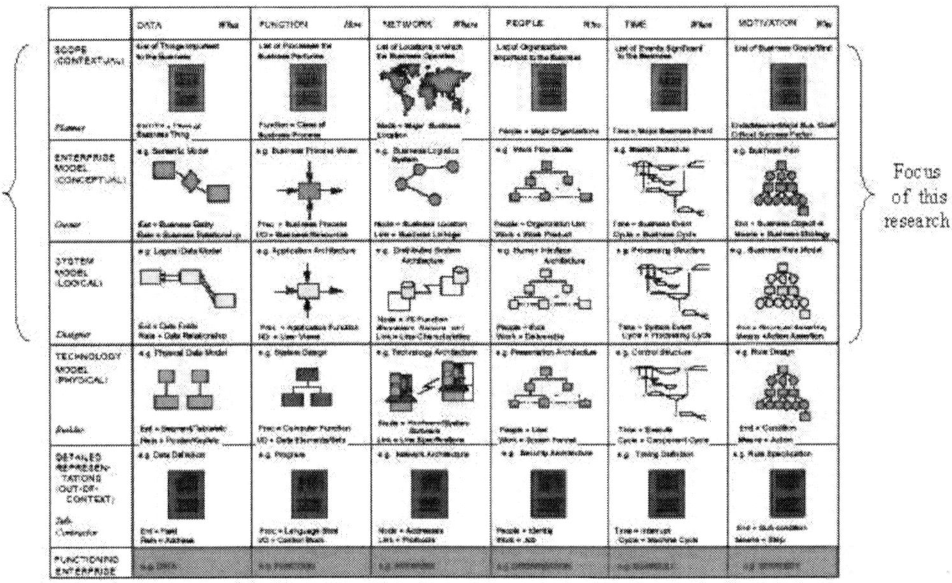

Figure 2. The Zachman Framework for Enterprise Architecture (Source: http://www.zifa.com, 2006)

Table 1. Zachman Framework at the Contextual, Conceptual and Logical Systems Architecture.

	What	How	Where	Who	When	Why	
SCOPE (Contextual) Planner	List of things important to the business	List of processes the business perform	List of location in which the business operate	List of people involved in the business	List of events significant to the business	List of business goals and objectives	SCOPE (Contextual) Planner
BUSINESS (Conceptual) Owner	Semantic Model	Business Process Model	Business Logistic Model	Work flow model	Master Schedule	Business Plan	BUSINESS (Conceptual) Owner
SYSTEM (Logical) Designer	Logical Data Model	Application Architecture	Distributed System Architecture	Human Interface Architecture	Processing Structure	Business Data Model	SYSTEM (Logical) Designer
TECHNOLOGY (Physical) Builder							
DETAILED PRESENTATION (Out-of-context view)							
FUNCTIONING ENTERPRISE - The prototype	**Data**	**Function**	**Network**	**People**	**Time**	**Motivation**	FUNCTIONING ENTERPRISE - The prototype

FORMULATION OF AN INTEGRATED FRAMEWORK FOR THE COLLABORATIVE DIGITAL LIBRARY

Collaborative digital libraries are constructed, collected and organized by a community of users and their functionalities support the information needs and uses of that community. Renda and Straccia (2004) viewed a digital library as a collaborative working and meeting space of people sharing common interests. Through a case study method, Zachman Framework is used as a basis to investigate the existing stakeholder's conditions and environment that would ensure the reception of a collaborative digital library for urban secondary schools use in Malaysia. In this digital library environment, students collaboratively build the digital library resources, which indirectly allow members of the community to be aware and be actively involved in local content development. The collaborative digital library would benefit both the students who would be the creator and publisher of digital project works and the teachers who would be given the experience of managing digital resources. The multi-method approach used in the case study, to ensure the consideration of all the aspects (dimensions) of a digital library system and the relationship of these dimensions in the framework used, has been reported elsewhere (Abdullah and Zainab, 2008). Findings were used to populate the Zachman Framework with contextual, conceptual, logical and module diagrams at every intersection between the columns (why, what, who, how, where, when) and the rows (scope, business model, system model). The framework abstracts the characteristics and features of the digital library based on the following six dimensions:

- Motivation factor, requiring the planner and owner to solicit answers to the "why" question, why there is need for the digital library? Why does the current business process need special handling such as those provided by the digital library?
- Data factor abstracts the "what" aspects of the digital library. What data that is currently handled by the stakeholders? What format would the data take in the digital library environment? What are the characteristics of the data used, processed, stored and presented or disseminated in terms of quality, accuracy, usability, description and organization?
- People factor looks at the "who" questions or the roles of people in the digital project environment. Who will be instructing? Who will be handling the data? Who will be reporting the collated or

processed data? Are the players in the digital library environment "ready" to participate and contribute to the digital library initiative? Are they able to do so?

- Function or Process factor defines the "how" of the activities in the digital environment. How will users search for data, how will they store the data? How will students write, present and submit their project report? How will the teachers ensure that the students know what is required? How do they grade the reports? How will they keep the reports submitted for the specified time required by the Malaysian Ministry of Education? How can the school library or resource centre accommodate these reports?
- Place or Networks looks at the "where" factor. Where will the digital library be located? Where will it be accessed by the stakeholders?
- Time looks at the "when" aspects of the digital library. When will submission of reports take place? This is useful for designing schedules, the processing and control architecture and timing systems.

The next section illustrates the use of Zachman Framework in design of the digital library, focusing on all six dimensions of the framework from three perspectives. Each of these dimensions is investigated from the perspective of the planner (Row 1), owner (Row 2) and designer (Row 3) of the digital library. These perspectives help ensure that everything relevant to the digital library enterprise is covered. The columns, comprising the six dimensions, are arranged so that the most important column or the focus of attention is presented first. At the end, the outcome would be in the form of listings and diagrams depicting the scope, business and system model of the digital library. Rows 4, 5 and 6 are beyond the scope of this chapter.

CASE ANALYSIS OF ZACHMAN FRAMEWORK FOR THE COLLABORATIVE DIGITAL LIBRARY

Motivation: why the digital library is needed

Why (Motivation) column of Zachman Framework extracts the motivation of the people that support the realization of the digital library. This reveals the reasons for creating the digital library, as well as the establishment of goals, objectives and business plan of the digital library. The authors felt motivation aspect (stakeholders' motivation) of the framework should be first populated and given the most importance. The case study revealed that the educational community is ready to collaboratively build the digital library as reflected by the following findings (Abdullah and Zainab, 2006):

- Students are "Internet ready", as indicated by (a) high home computer ownership as such they are ready to utilize the digital library; (b) high Internet penetration either at home, school, cyber cafes or friend's houses; (c) a high number of students either have 3-4 years or more than 5 years experience in computer usage; (d) students regularly go online, between either every alternate days or everyday.
- Students are "digital ready" as indicated by their awareness, experience in using and preferring digital sources. The survey indicates that all students know how to word process; they know how to prepare slide presentations or draw using the computer, edit images, create multimedia and scan images, create web pages, database or undertake simple programming. This results show that they are aware and competent in handling digital resources, which is necessary when using digital library.
- Students are also moderately "Web ready" as they know how to use the web. Although most had no formal training, they learn how to use the Web by self-teaching, from books and people. The students also learn how to find sources on their own, from their parents and siblings or from their classmates and friends.
- Students are ready to collaboratively develop digital resources as they indicated sharing the resources they create or found with their friends either by e-mailing the URLs of websites, informing others through chat room or social networks or creating links to websites.
- Some students are "ready web publishers", as many of them either maintain a group web page, have their own personal web page or have been creating page pages for others. The results also show that the students have experience in creating digital resource over the Internet using webpage creator tools or HTML to develop their sites.
- Students do use the Internet for school related assignments or as a major source for their school project. Students sampled highly use the Internet resources to get information for the following subjects,

History, Science and Geography. The results indicate that students believe that the Internet helps them with their school work.

- All students feel that there is a need for digital library of local information and feel that this would definitely benefit them. The results indicated very significant correlation between positive perception of the digital library with high Internet use, length of Internet experience, accessibility of Internet from home high self ratings of Internet skills.
- Teachers in the case study see the value of digital resources and online publishing for their students. They expressed the willingness to play the role as a facilitator in the digital library environment. They were keen on the digital library because students could contribute original works to be shared with other students, especially where some local contents are not available in textbooks. They believed students would be more careful in preparing their project report if they make it available to wider audience in the digital library environment.
- Teachers felt that a digital library would solve problem of storage and retrieval of reports submitted since project reports kept in resource room could only be retrieved by class name and level, and difficult to be retrieved by subject or topics of report or by specific student's name.
- The school's infrastructural facilities are ready to support a digital library as students can gain access to computers at self-accessed learning centres, especially for students and teachers who do not have computers or Internet access at their homes.

The above findings indicated that the readiness factors serve as motivating indicators, goals and objectives that support the plan for developing the digital library. Figure 3 presents the motivating factors that support the plan of the collaborative digital library.

The findings of the case study were plugged into the first three rows of the motivational aspect of Zachman Framework. In Row 1, the Planner's goals and objectives are defined in the form of vision statement that provides the strategic direction for the digital library. The digital library will support secondary students' information needs in conducting research projects through project-based learning (PBL). In PBL, students interpret, analyze, synthesize, generate, and evaluate information about a topic, collaborate with others, and produce a report (Blumfeld et al, 1991). To support students in these types of activities, a full complement of tools is needed to meet the unique needs of learners, and Internet technologies such as digital libraries have the affordances to support students in these activities. Based on this premise, as well as building from various illustrations of digital library initiatives' vision statement, the planner establishes the vision of the collaborative digital library to populate Row 1 of the Motivation column. The planner's vision of the digital library is as follows:

"The collaborative digital library should enable secondary students conducting history[1] school projects to access the information they need any time and any where, in a friendly, efficient and effective way, by overcoming the barriers o distance and language. The digital library should enable students to collaboratively contribute resources as the digital library is seen as a growing repository on Malaysian local history for education".

With the vision in mind, the planner establishes the following goals for developing and implementing the collaborative digital library:

a. the development of local historical resources;

b. provision of resources for lifelong learning;

c. provision of round-the clock access; and

d. development of community of users.

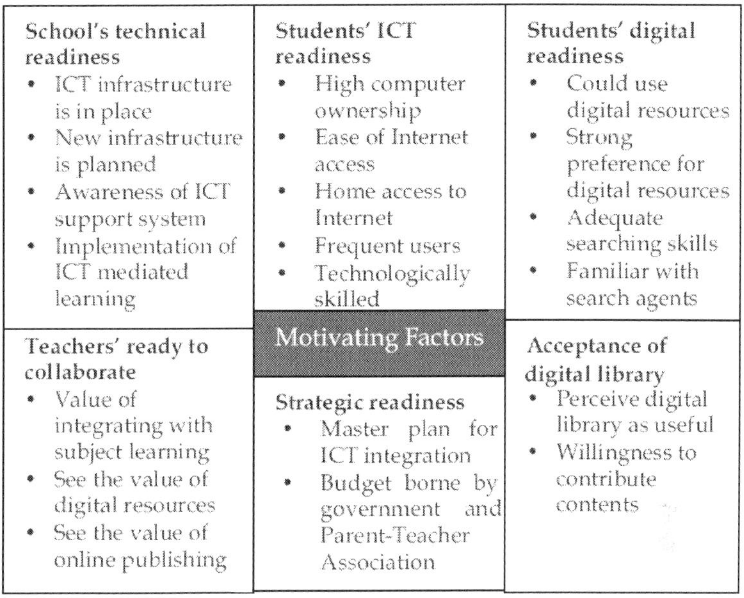

Figure 3. Motivating Factors that Support the Plan for Realization of the Digital Library

In this capacity, it establishes "a digital library service environment" – that is, a networked, online information space in which students can discover, locate, acquire access to and, increasingly, use information. The objective of the digital library is therefore to provide a learning environment and resources network for history education which is:

a. designed to meet the information needs of learners, in both individual and collaborative settings (enable the creation, organisation and maintaining of local history resources);

b. constructed to enable use of a broad array of materials for local history learning, primarily in digital format submitted by the educational community themselves (they themselves become resource providers); and

c. managed actively to promote reliable access anytime - anywhere to quality collections and services (provided over the Internet), available both within and outside the network.

Row 2 of the Motivation column identifies the owners' business plan that is the approach to use the collaborative digital library. The digital library is modelled to focus on serving students information needs in conducting research projects. As such, in the implementation of this digital library project, the use of the online resources would be an integral part of history projects-based learning activities. The digital library may move the student community towards an emerging digital resources and the submission of reports in the electronic form is therefore feasible. The implementation of the business plan (Figure 4) is consistent with the Ministry of Education's implementation and evaluation of History project, which will make the accomplishment of the goals and objectives feasible. The teachers on the other hand will be given the opportunity to validate the quality of submissions to maintain the quality of the digital library, grade the report online and add links to resources found on the Internet.

Row 3 spells out the the designer's perspectives which expressed the motivation of the digital library in the form of behavioural objectives. The objectives of the collaborative DL from the designer's perspectives are to

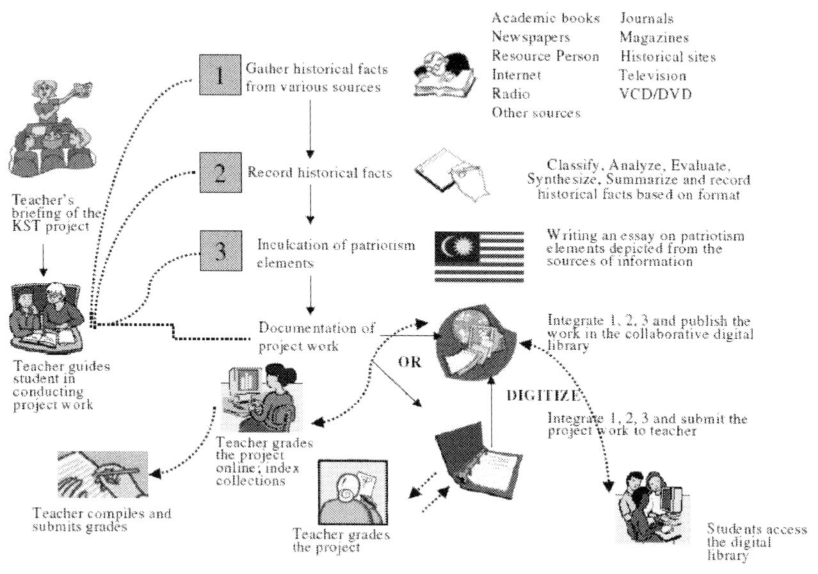

Figure 4. The Business Plan to Use the Collaborative Digital Library (Owner's View of Motivation)

a. Enable students to search and browse the digital library resources through various access points regarding the topics they are exploring

b. Allow the students to sequence and organize their project reports in various styles, construct references and append digital objects or pictures to their report.

c. Provide the students with the experience of publishing their project report, allow teachers grade and their friends to view the repot. The motivation for this is to satisfy their innate need to share their work, so that their peers can give comments for improvements before the report is finally submitted.

d. Allow teachers to check the suitability of submissions, maintain quality of contents of the digital library and grade the submissions.

e. Allow teachers and students to provide metadata for resources submitted to the digital library. A metadata schema (Dublin core) will be applied for this purpose.

f. Enable students and user groups to register as members to login and submit and describe resources

g. Allow users to submit feedback or submit useful links to other resources in the Internet.

h. Guide and assist users in using the digital library functions and services.

i. Allow authorized users to add, modify or delete submitted resources to the digital library.

These behavioural objectives of the digital library would assist the designer (Row 3) in developing the required digital library. The motivation and objective statements subsequently assist in the development of the user requirement and detailed definitions of the digital library services required in the Function (How) column of the Framework.

Data: what resources constitute the digital library

What (Data) column of the Zachman Framework describes the digital library resources students used to fulfill their research needs. The data component, at the macro level identifies the information resources included or covered in the collaborative digital library, and at the micro level, concerned the collections, quality, accuracy, usability, description and organisation of the resources in the digital library. Findings from the case study revealed that the students and teachers emphasized the needs for contents to be "clear, accurate, adequate, organised, valid, reliable, informative and resourceful"

To cater for students' information needs, in Row 1, the planner describes the three main categories of resources (Figure 5), without policy-controlled access. The types of resources are (a) resources that are born digital; (b) digitised resources or digital proxies for physical items; and (c) Links to other resources relevant to the domain focus of the digital libbrary. The digital library collections incorporates not only digital resources in different media types such as text, images, web documents, audio and video, but also in different formats with different levels of content quality and metadata.

Row 2 of the Data column is a contiguous model of the resources expressed in terms of domain focus and topics seen by the owners of the digital library. History has been chosen as the domain of the digital library test-bed based on the survey findings that indicated the students surveyed mainly use Internet resources to get information for their History project.

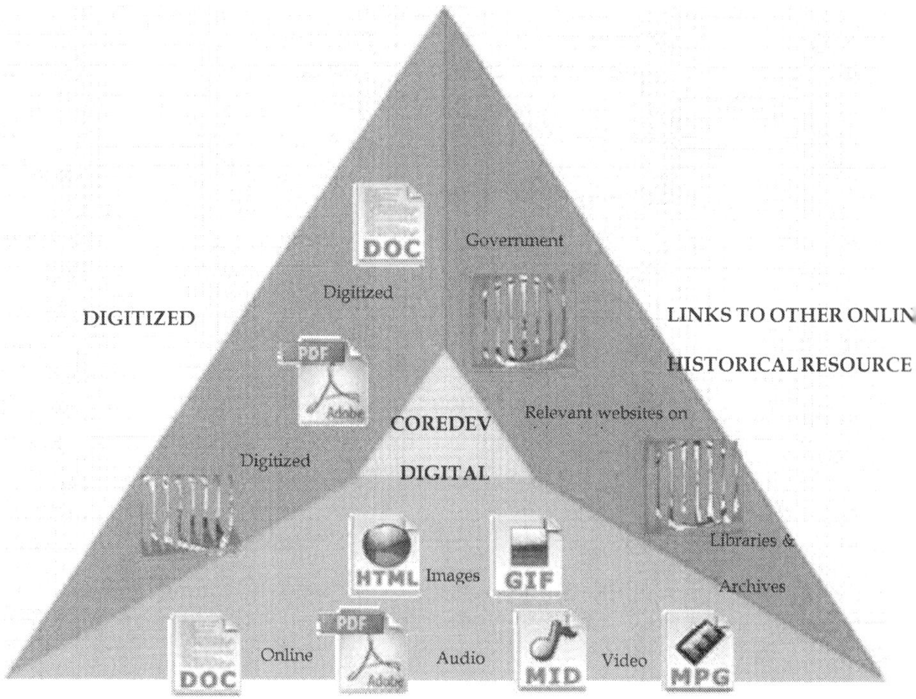

Figure 5. The Digital Library Resources in Various Media Types and Format (Planner's View of Data)

Case findings revealed that the project reports are typically made available in the form of collections, which refers to groups of resources organised around three themes or topic namely prominent personalities, historical events and historical buildings. Figure 6 presents the semantic description of the domain focus, contents, content criteria and scope of the collaborative digital library, which populates the Data component of the Zachman Framework. The stakeholders' needs for contents to be "clear, accurate, adequate, organised, valid, reliable, informative and resourceful" are therefore used as a set of general guidelines or selection criteria of resources accepted for submission.

From the designer's perspective (Row 3), the data of the digital library is expressed as table definition for the digital library data (comprising digital objects data and metadata, user information, annotation and static information pages) and metadata profile for the digital object resource description (comprising administrative, technical and descriptive metadata). Administrative metadata is created by the author, technical metadata is automatically- generated and descriptive metadata is assigned by the content access provider (human indexer). The descriptive metadata schema used for the object data description is the Dublin Core (DC) Metadata. The digital library has altogether 16 metadata elements and incorporates DC's 14 out of 15 elements, namely title, creator, subject, description, publisher, contributor, date, type, format, identifier, language, relation, coverage and rights. The DC source metadata element is not used. Two other elements incorporated are Collection and Ranking metadata.

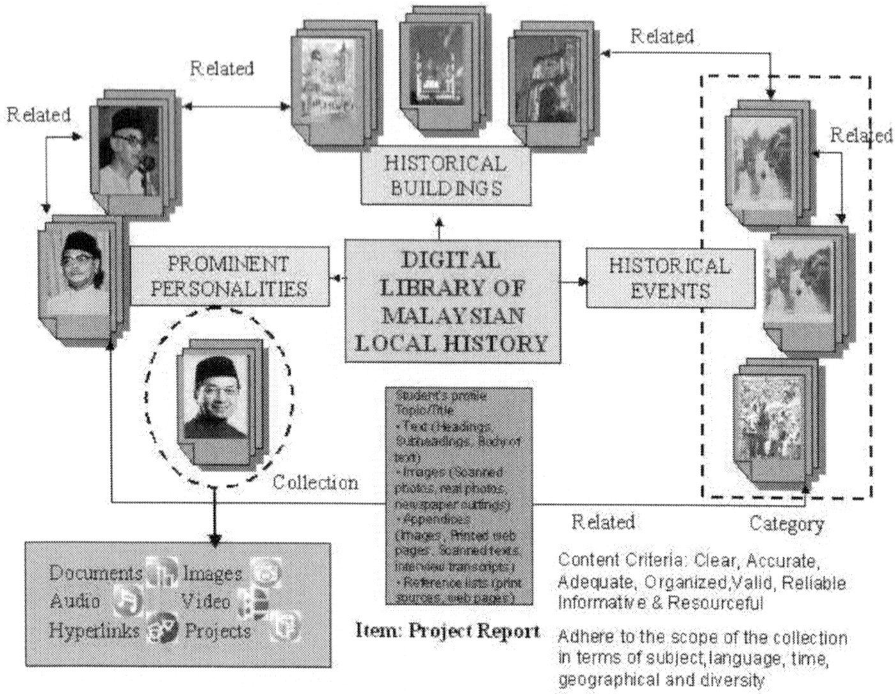

Figure 6. Domain Focus, Contents, Content Criteria and Scope of the Collaborative Digital Library (Owner's View of Data)

People: who interacts with / within the digital library

Who (People) column represents the stakeholders or the people within the digital library enterprise to which the digital library assigns responsibility for work. Thus, this component concerns the identification of the digital library users, their information needs, their usage of the Internet and online digital resources and their roles in the enterprise. The design of the enterprise has to do with the allocation of work and the structure of authority and responsibility. This column also deals with human-machine interfaces and relationships between the people and the work they perform.

In Row 1, the planner identifies the audience and the digital library organization. There are three types of audiences within the collaborative digital library enterprise, categorized as partners, guests and affiliate members (Figure 7). The planner identified these groups of people in the form of digital library organisational structure.

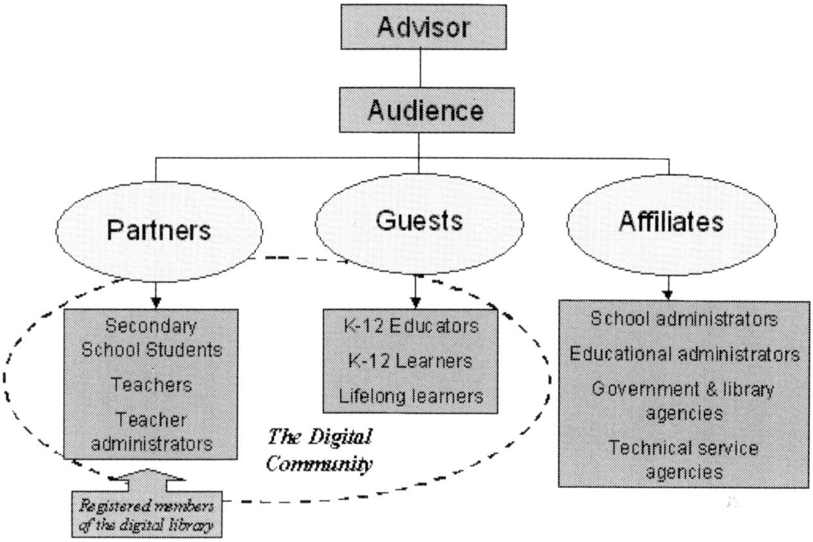

Figure7. The Digital Library Organisational Structure (Planner's View of People)

From the owner's perspective, Row 2 of the People Column illustrates the four main classes of people or actors and their respective roles in the collaborative DL. In this Consumer – Content Provider – Content Manager – Administrator model, each class of actors represents a particular generic role. The digital community follows certain rules and their members play different roles, as consumers, content developers or providers, content access providers and content manager (Figure 8).

In Row 3, the designer fleshed out the interaction between actors and technology into a rich picture linked to the functional requirements (Figure 9). Here, the digital library community includes people as well as computers, agents, network connections, files and operating systems, user interfaces, communication links, and protocols, which either use or support the digital library services. The communities of autonomous agents and computers instantiate functions upon requests by the actors of the digital library. To operate, these agents and computers need structures of vocabulary and protocols. They act by sending streams of queries and retrieving streams of results. The digital library system uses the three-tier client-server architecture. As depicted in Figure 9, the client tier comprises computers with web browsers such as Internet Explorer

(4.0 or above), Netscape Navigator, Mozilla Firefox and Opera. User interfaces are provided for clients to process their application and manipulate their data. All application programmes resid

e in the middle-tier (web server). The web server processes the request from the client and then returns required result in web page format. It processes data request by linking to a database server (such as authenticating and validating users that login into the system). It is also linked to transaction server, especially when clients are uploading files to the web server. The third tier consists of the database server and transaction for maintaining data records.

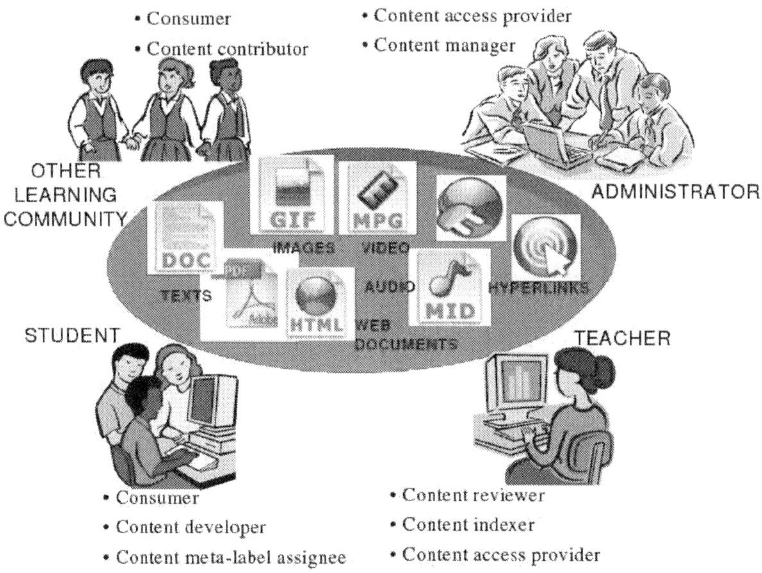

Figure 8. The Digital Library Actor-Roles Diagram (Owner's View of People)

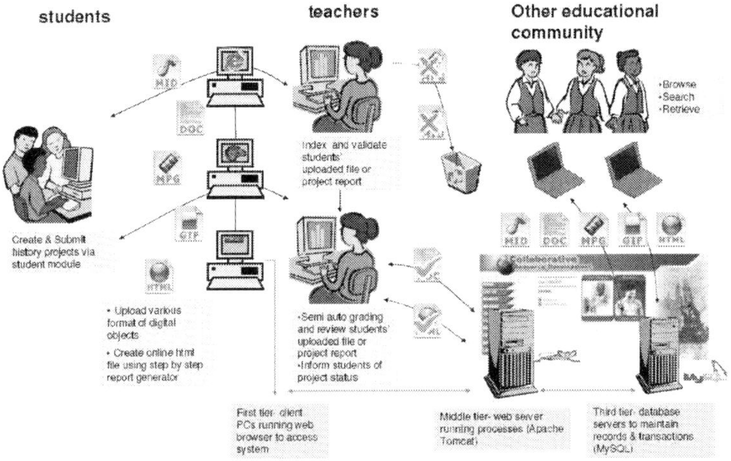

Figure 9. Actors and Their Roles Depicted in the Digital Library Three-Tier Client-Server Architecture (Designer's View of People)

1.1 Function: what happens in the digital library?

How (functions) defines the functions or activities the digital library enterprise is concerned about relative to each perspective. In Row 1, the planner describes the students' research activities that take place, which encompass the entire information seeking process (from recognizing the need for information to finding, using and presenting it) and the submission and evaluation of the information in the form of project report. This is presented in the form of rich pictures Basically, the students do solitary information seeking, have spontaneous interactions with other people such as parents, siblings and friends and ask for help, and work with information in a group. The description of the activities when conducting history projects are then transformed into the online activities the students and teachers would be able to perform in the collaborative digital library. Figure 10 presents the workflows and processes the collaborative digital library enterprise should conduct. These processes are also in line with the owner's plan to use the digital library for school project (Row 2 of Motivation). The function component refers to the activities students perform in their research, such as choosing topic, searching for information, organising resources, writing, presenting, submitting and teachers grading of project work.

Using data from analysis of the activities culled from the research (Row 1 of Function), formulation of behavioural objectives of the digital library (Row 2 of Motivation) as well as from the analysis of digital library functional requirements (Row 3 of Motivation), the planner develops the user requirement expressed in terms of functions and present it as services in a contiguous structured chart, The structured chart is comprehensible to the owner as the conceptual model of the digital library services (Figure 11). This structured chart populates Row 2 of the Function Column and describes the process of translating the objectives of the digital library enterprise into successively more detailed definitions of its services. Feedback from the stakeholders on the potential features of a service and digital library design implication derived from the analysis of the case study have helped to ascertain the main features required by the collaborative digital library.

In Row 3, the designer portrays the digital services in terms of data transforming processes, described exclusively in terms of definition of programme modules and how they interact with each other. The three system modules, namely administrators (including teachers), students and guests, provide different access types for different level of users. Along with this are specific definitions of security requirements, in terms of who (which role) is permitted access to what function, in the form of structured charts and detailed description of the modules menu.

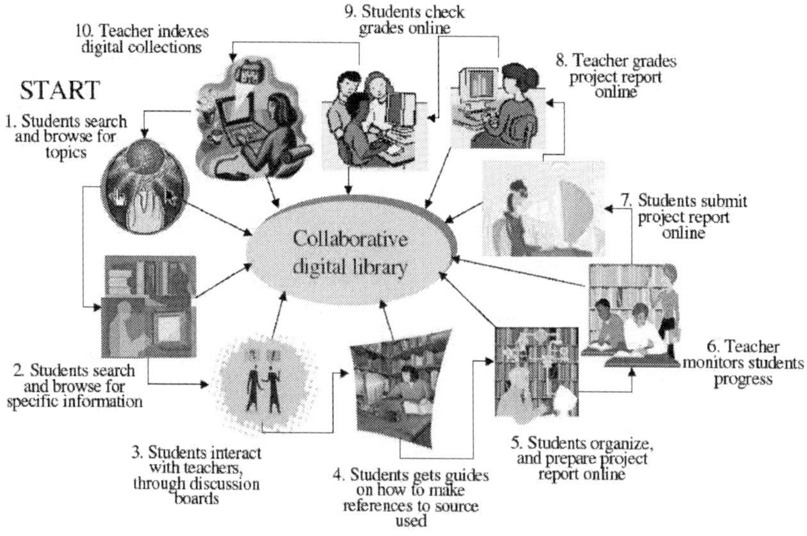

Figure 10. Activities Performed in the Collaborative Digital Library (Planner's View of Function)

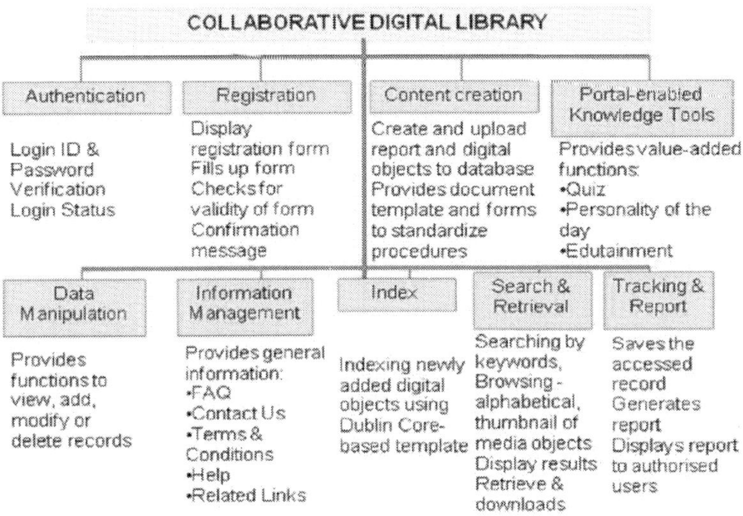

Figure 11. Conceptual Model of Services in the Digital Library (Owner's View of Function)

Network: where can one access the digital library

Network (Where) shows the sites or geographical locations and the interconnections between activities within the digital library enterprise. It illustrates the network-related aspect of the digital libraries in terms of the physical locations of members in the digital library which spread over a geographical area. The planner provides the big picture of the digital library as a centralized system with the control for the whole structure at the Faculty of Computer Science and Information Technology University of Malaya (FCSIT UM) as the developer of the digital library system (Figure 12). FCSIT UM group manages the centralized database server. School A is the content collaborator and joint owner of the system and other potential future collaborators such as School B, Education Departments, Ministry of Education, as well as other repositories, would be able to utilise the application server running locally to fetch the required data from the database server.

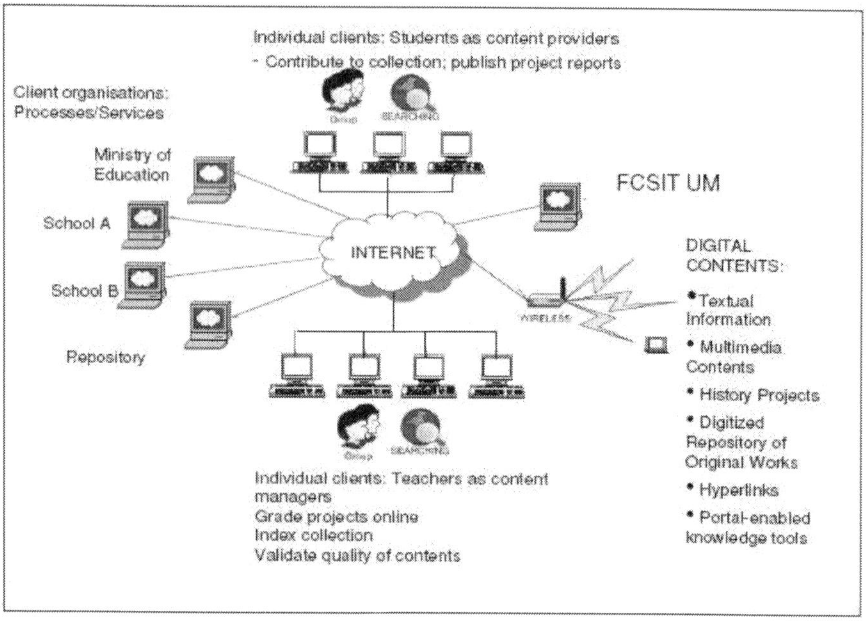

Figure 12. The Physical Network of the Collaborative Digital Library (Planner's View of Network)

The owner is interested in the conceptual model of "Where" which includes the location of access and place where the primary stakeholders, namely the students and teachers use the digital library (Figure 13). It illustrates the collaborative digital library deployment expressed in term of location of access and computing facilities and network. The school community may access the collaborative digital library system from any 10 locations in the school, as all computers there are connected to the network.

From the designer's perspective, the Network Column presents the logical model of the network component of the collaborative digital library which depicts the types of systems facilities and controlling software at the nodes and lines such as processors/operating systems, database and lines/line operation systems. The notional distributed systems architecture (Figure 14) shows servers supporting the digital library services served from the regional (FCSIT) and local data center environment to the school's three primary locations of access. It is referred to as a notional architecture since the extent of the ability to remotely serve specific applications in both the baseline state and the target state remains to be established

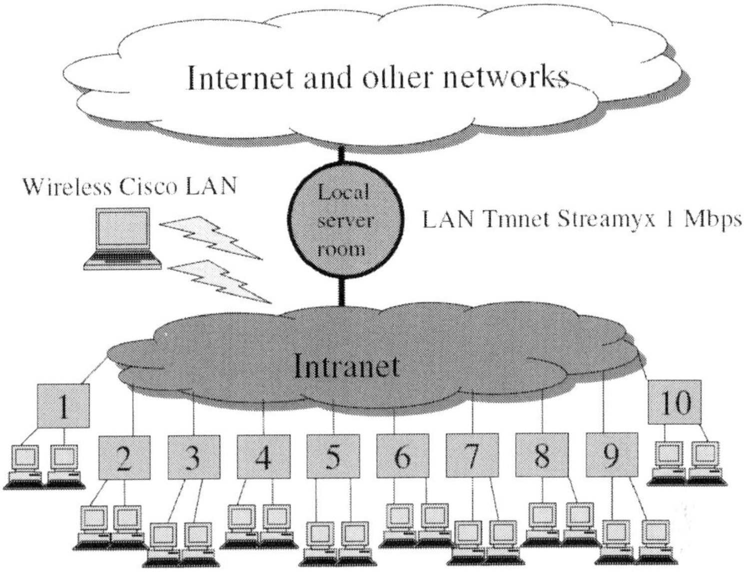

1	Self-Access Learning Centre	10	6	Staff room 1	4
2	School Resource Centre	5	7	Staff room 2	4
3	Computer Laboratory	40	8	Department Heads' Office	4
4	Information Technology Laboratory	40	9	Subject Panels' Room	4
5	Language Laboratory	20	10	Main Office	10

Figure 13. The School's Network Diagram Positioning the Location of Access for the Digital Library (Owner's View of Network)

Time: when can one use the digital library (when do things happen)

The last column, "When" represents time, or the events to which the digital library responds in relation to time. This is useful for designing schedules, the processing architecture, the control architecture and timing systems. It is difficult to describe or address this column in isolation from the others, especially Column 2 (Process). At the strategic level, the planner describes Time as the business cycle and overall business events. As has been delineated in the digital library goals and objectives (Motivation Column), the digital library provides round-the-clock access. As the Internet is a 24/7 medium, the digital library is available 24 hours a day, 7 days a week.

In the detailed model of owner's perspective, the Time Column defines when activities or processes are to happen. Based on the findings of the case study regarding the school's approach in using the digital library, the chronology of events (such as teacher's notification and requirement of the project, students choose topic, gather information, create report, obtain teacher's feedback, edit and submit report) indicating the processes that take place in the digital library environment populates owner's view of the Time Column. The designer defines the business events or the processes in the digital library, which cause specific data transformations and entity state changes to take place (Table 2). The business events populate the designer's view of the Time Column of the Zachman Framework used.

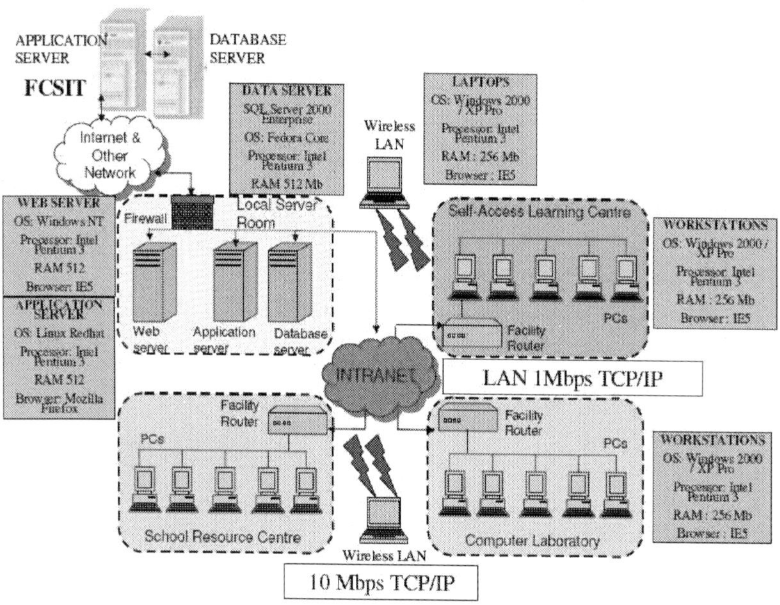

Figure 14. The Digital Library Notional Distributed Systems Architecture (Designer's View of Nework)

Table 2. Business Events in the Digital Library (Designer's View of Time)

The Process	Data transformations and entity state changes to take place.
Students register	Students receive automatically generated e-mail notifying membership of the digital library
Students create and submit report	Teachers and Administrators receive automatically generated e-mail notification indicating a new report has been submitted and ready to be viewed, graded or indexed.
Students create and submit report	Students receive automatically generated e-mail notification indicating that they have successfully submitted their project report.
Administrator registers teachers	Teachers receive automatically generated e-mail notification which indicates their User ID and Password.
Teacher evaluate and grade report	Students receive automatically generated e-mail notification indicating their projects have been evaluated.

CONCLUSION

This chapter has provided a detailed mapping between the first three layers of Zachman Framework for Enterprise Architecture and the concepts utilized in formulating the requirements and design of a digital library, based on a case study and analysis on building a collaborative digital library to meet the needs of the stakeholders. It has also illustrated the possibility of using the Zachman Framework as an instrument for requirements analysis and evaluation in digital development. The framework highlights the need to involve all possible stakeholders in the development of the architecture, not just the enterprise architects and developers, to ensure that it meets their needs and uses.

The perspectives and artifacts established from the framework have helped ensure that everything relevant to the digital library enterprise is covered. Table 3 shows the mapping of the Zachman Framework perspectives and dimensions with the collaborative digital library deliverables/aspects. The planner's perspective reflects the context that establishes the list of relevant constituents that must be accounted for in the descriptive representation for the other perspectives (owner and designer). The descriptive representation of owner's perspective reflects the usage characteristics of the digital library, what the owner is going to do with it and how they will use it once they get it in their possession. The descriptive representation of designer's perspective forms the basis for the design of the digital library system, as well as the features for manipulating the tangible aspects of the digital library.

Table 3. Mapping the Digital Library Deliverables to the Zachman Framework Perspectives and Dimensions

DIMENSIONS	ZACHMAN FRAMEWORK	THE DIGITAL LIBRARY
LEVEL 1: Objectives/Scope – Planner's View		
Motivation (Why)	Identify and list goals and objectives - requirements analysis based on identified objectives	Motivating factors diagramme Vision statement List of DL goal List of DL objectives
Data (What)	Identify & list features /data important to the repositories using needs survey from stack holders	**DL resources in various media types and format**
People (Who)	Identify & list all stack holders and their roles in handing & processing data - Roles analysis	DL organisational structure
Function (How)	Identify & list processes the data performs - stack holder's information use survey	Activities students perform when conducting history project in the DL
Location (Where)	Identify & list locations where the enterprise operates - information flow survey	The physical network of the DL
Time (When)	Identify & list business events cycles - events use analysis	Access to the DL (24/7)
LEVEL 2: Business Conceptual Model – Owner's View		
Motivation (Why)	Business Plan Business plan - flow diagram / rich picture	Rich picture showing owners' approach to use the DL List of perceived benefits to use the DL
Data (What)	Entity relationships diagrams - rich picture	DL Subject scope, collection and resource criteria
People (Who)	Organisation charts, roles, set of skills & security issues - box charts / rich picture	DL actor-roles diagram
Function (How)	Business process model - flow diagram / rich picture	Conceptual model of services in the DL
Location (Where)	Logistics network - nodes and links	Owner's network diagram positioning the location of access for the DL
Time (When)	Business master schedules -rich picture	Chronology of events in the DL environment
LEVEL 3: System Model – Designer's View		
Motivation (Why)	Business rule model	Behavioural objectives as DL mandatory functional requirements
Data (What)	Data model - entity diagrams	Table definitions for the DL data; metadata profile for the digital objects resource description
People (Who)	System interfaces architecture indicating roles, data, access	Actors and their roles depicted in the DL's three-tier client-server architecture
Function (How)	Data flow diagram showing application of data in the architecture	Structured chart for programme modules; users' menu
Location (Where)	Diagram indicating how data is distributed	The DL notional distributed systems architecture
Time (When)	Process structure with dependency diagrams, entity life history	Business events in the DL

Using Zachman Framework as the approach to design a digital library has contributed to the field of Enterprise Architecture by highlighting the fact that fusion of information technology with business is important and these two aspects should be addressed together in organizations. It has also contributed to another dimension of a framework for digital library research and development and "a structured vision for the development of new ideas" (Soergel, 2002). The collaborative digital library adheres to Soergel's guiding principles and ten themes for digital library research and development, as well as incorporates the dimensions of others' framework but instead of listing them as requirements or ticking against a checklist, the authors have embedded the requirements in a system's architectural framework and present them more systematically, taking into account the following details, aligned with the emerging perspective of the Enterprise Architecture:

a. The vision, goals, objectives, business plans and the functional requirements of the digital library;

b. The types of resources, domain focus, collection and data definition of the digital library;

c. The stakeholders and users, their roles and functional roles in the digital library;

d. The activities users perform, service conceptual model and the digital library programme modules

e. The location of access, the network diagram and the notional distributed system architecture;

f. The availability, business cycle and overall business events

In alignment with the vision already expressed by the DLF (2005), the authors felt that digital library developers should "get out of the box" and give more attention to the development of conceptual frameworks giving preference to scopes, goals requirements and processes, in the sense as those concepts are already common in the classic Enterprise Architecture processes and Zachman Framework for Enterprise Architecture can be a very simple comprehensive reference for this. Perhaps is time for the digital library researchers and practitioners to recognize that the focus of the digital library should move from the perspective of the engineer, who are responsible for systems design, to the perspective of the architect who prepares, plans and develops specifications, that bridge the gap between the systems (that the engineers design) and what the community needs. Although this chapter defines the design process and constructs necessary for the development of a collaborative educational digital library for secondary schools with a particular focus in Malaysia, illustration of the detailed mapping between the first three layers of Zachman cells and the dimensions utilized in formulating the requirements and design of the digital library can facilitate design transferability so that it could also be applied in another country setting.

REFERENCES

1. Abdullah, A & Zainab A.N. (2006). Ascertaining factors motivating use of digital libraries and formulating user requirements using Zachman Framework. *Malaysian Journal of Library Information Science* Vol.11 No 2: 21-40, ISSN 1394-6234
2. Abdullah, A. & Zainab, A.N. (2008). The digital library as an enterprise: The Zachman approach. *The Electronic Library*,Vol. 26, No. 4, 446-467, ISSN 0264-0473.
3. Blumenfeld, C., Soloway, E., Marx, R. W., Krajcik, J. S., Guzdial, M., and Palincsar, A. (1991). Motivating project-based learning: Sustaining the doing, supporting the learning. *Educational Psychologist.* Vol. 26, No. 3&4, 369-398, ISSN 0046-1520
4. Bolton, G. (2004). What is Enterprise Architecture? Available at http://www.eacommunity.com/resources/whitepapers.asp
5. Borbinha, J. (2007). It is the time for the digital library to meet the enterprise architecture, *Proceedings of the 10th International Conference on Asian Digital Libraries: looking back 10 years and forging new frontiers, ICADL2007.* Lecture Notes in Computer Science, pp. 176-185, ISBN 978-3-540-77093-0, Hanoi, Vietnam December 10-13, 2007, Springer-Verlag, Berlin
6. Castelli, D. & Fox, E.A. (2007). Series of workshops on Digital Library Foundations. *D-Lib Magazine*, 13, 9/10 (Sept/Oct 2007), ISSN 1082-9873, Available at: http://www.dlib.org/dlib september07castelli/09castelli.html
7. Fuhr, N.; Hansen, P.; Mabe, M.; Micsik, A. & Solvberg, T. (2001). Digital libraries: A generic classification and evaluation scheme. In *Proceedings of the 5th European Conference on Research and Advanced Technology for Digital Libraries, ECDL2001*. Lecture Notes in Computer Science, pp. 187-199, ISSN 0302-9743, Darmstadt, Germany, September 4-9 2001, Springer-Verlag, Berlin
8. Digital Library Federation. (2005). DLF service framework for digital libraries: A progress report for the DLF steering committee.

Available at: http://www.diglib.org/architectures/serviceframe/dlfserviceframe1.htm

9. Gonçalves, M. A.; Fox, E. A.; Watson, L. T. & Kipp, N. A. (2004). Streams, structures, spaces, scenarios, societies (5s): A formal model for digital libraries. *ACM Trans. Inf. Syst.* Vol. 22, No. 2 (Apr. 2004), 187-199, 270-3122163, Springer-Verlag, Berlin
10. Gladney, H.; Fox, E. A.; Ahmed, A.; Ashany, R.; Belkin, N. J. & Zemankova, M. (1994). Digital library: Gross structure and requirements: Report from a March 1994 Workshop. In *Proc. 1st Annual Conf. on the Theory and Practice of Digital Libraries,* IBM Research Report RJ 9840
11. Kahn, R. & Wilensky R. (1995). *A framework for distributed digital object services.* Available at: http://www.cnri.reston.va.us/cstr/arch/k-w.html
12. Lavoie, B, Henry G. and Dempsey. L. 2006. A service framework for libraries. *D-Lib Magazine,.* Vol. 12, no 7/8, (July/Aug, 2006) ISSN 1082-9873. Available at: http://www.dlib.org/dlib/july06/lavoie/07lavoie.html
13. Levy, D. & Marshal, C. (1995). Going digital: A look at assumptions underlying digital libraries. *Communications of the ACM.* Vol. 38 , No. 4, 77-84, ISSN 0001-0782
14. Marchionini, G. & Fox, E. A. (1999). Progress toward digital libraries: Augmentation through integration. *Information Processing & Management.* Vol. 35, No. 3, 219-225, ISSN 0306-4573
15. McGovern, J.; Ambler, S.W.; Stevens, M.; Linn, J.; Sharan, V. & Jo, E. (2003). *The practical guide to Enterprise Architecture.* ISBN-10: 0131412752, Englewood Cliffs, N.J. : Prentice-Hall
16. Moen, W.E. & McClure, C.R. (1997). *An evaluation of the U.S. Government's implementation of the Government Information Locator Service (GILS).* Available at: http://www.unt.edu/slis/research/gilseval/gilsdocs.htm
17. Pereira, C.M. and Sousa, P. (2004). A method to define an Enterprise Architecture using the Zachman Framework. *ACM Symposium on Applied Computing 2004,* Nicosia, Cyprus, March 14-17.
18. Renda, E. M. and Straccia, U. (2004). A personalized collaborative digital library environment: a model and an application.

Information Processing and Management. Vol. 14, No.1, 5-21, ISSN *0306-4573*

19. Sandusky, R. J. (2002) Digital library attributes: Framing usability research. In Blandford, A. and Buchanan, G. (eds.) *Proceeding Workshop on Usability of Digital Libraries at JCDL'02.* (p. 35-38)
20. Saracevic, T. & Covi, L. (2000). Challenges for digital library evaluation. In Kraft, D. H. (ed.), ASIS 2000: *Proceedings of the 63rd ASIS Annual Meeting,* Vol. 37 (p. 341-350). Medford, NJ: Information Today
21. Schauble, P. & Smeaton, A.F. (1998). *Summary report of the series of joint NSF-EU working groups on future directions for digital library research: An international research agenda for digital libraries.* Available at: http://www.ercim.org/ publication/ws-proceedings
22. Soergel, D. (2002). A framework for digital library research: Broadening the vision. *DLib Magazine.* Vol. 8, No. 12 (Dec 2002), ISSN 1082-9873. Available at: http://www.dlib.org/dlib/december02/soergel/ 12soergel.html.
23. Zachman, J. A. (2002). *Zachman International Enterprise Architecture.* Available at: http://www.zachmaninternational.com/Default.htm

INDEX